Clinical Nutrition for Cancer Patients

Clinical Nutrition for Cancer Patients

Editor

Vera Mazurak

MDPI

Basel • Beijing • Wuhan • Barcelona • Belgrade • Novi Sad • Cluj • Manchester

Editor
Vera Mazurak
University of Alberta
Edmonton
Canada

Editorial Office
MDPI
St. Alban-Anlage 66
4052 Basel, Switzerland

This is a reprint of articles from the Special Issue published online in the open access journal *Nutrients* (ISSN 2072-6643) (available at: https://www.mdpi.com/journal/nutrients/special_issues/clinical_nutrition_cancer).

For citation purposes, cite each article independently as indicated on the article page online and as indicated below:

Lastname, A.A.; Lastname, B.B. Article Title. *Journal Name* **Year**, *Volume Number*, Page Range.

ISBN 978-3-7258-0533-4 (Hbk)
ISBN 978-3-7258-0534-1 (PDF)
doi.org/10.3390/books978-3-7258-0534-1

Contents

Preface

This special issue highlights research that spans the breadth from diagnosis through treatment, and beyond, addressing improvements in assessment and ways to address the nutritional challenges faced by patients with a cancer diagnosis. Importantly, emerging research for interventions to circumvent malnutrition in this complex disease are presented.

Vera Mazurak
Editor

nutrients

MDPI

Review

The Neuroendocrine Neoplasms of the Digestive Tract: Diagnosis, Treatment and Nutrition

Jakub Pobłocki [1], Anna Jasińska [2], Anhelli Syrenicz [1], Elżbieta Andrysiak-Mamos [1] and Małgorzata Szczuko [2,*]

[1] Department of Endocrinology, Metabolic Diseases and Internal Diseases, Pomeranian Medical University in Szczecin, Unii Lubelskiej 1str, 70-252 Szczecin, Poland; jakub.poblocki@pum.edu.pl (J.P.); klinendo@pum.edu.pl (A.S.); Elamamos@o2.pl (E.A.-M.)

[2] Department of Human Nutrition and Metabolomic, Pomeranian Medical University in Szczecin, Broniewskiego 24 str, 71-460 Szczecin, Poland; a.jas@wp.pl

* Correspondence: malgorzata.szczuko@pum.edu.pl; Tel.: +48-91-4414810; Fax: +48-91-441-4807

Received: 9 April 2020; Accepted: 13 May 2020; Published: 15 May 2020

Abstract: Nuroendocrine neoplasms (NENs) are a group of rare neoplasms originating from dispersed neuroendocrine cells, mainly of the digestive and respiratory tract, showing characteristic histology and immunoprofile contributing to classification of NENs. Some NENs have the ability to produce biogenic amines and peptide hormones, which may be associated with clinical syndromes like, e.g., the carcinoid syndrome caused by unmetabolized overproduced serotonin, hypoglycemic syndrome in case of insulinoma, or Zollinger-Ellison syndrome accompanying gastrinoma. Diagnostics for these include ultrasound with endoscopic ultrasound (EUS), computed tomography (CT), magnetic resonance imaging (MRI), and positron-emission tomography/computed tomography (PET/CT). Different nuclear medicine procedures can also be used, like somatostatin analogues scintigraphy (SRS) and 68Ga-Dota-Peptide PET/CT, as well as biochemical methods to determine the level of general neuroendocrine markers, such as chromogranin A (CgA), 5-hydroxyindolacetic acid (5-HIAA), synaptopfysin and cell type-specific peptide hormones, and neurotransmitters like gastrin, insulin, serotonin, and histamine. NENs influence the whole organism by modulating metabolism. The treatment options for neuroendocrine neoplasms include surgery, somatostatin analogue therapy, radionuclide therapy, chemotherapy, molecular targeted therapies, alpha-interferon therapy, and inhibitors of serotonin production. In the case of hypersensitivity to biogenic amines, a diet that limits the main sources of amines should be used. The symptoms are usually connected with histamine, tyramine and putrescine. Exogenic sources of histamine are products that take a long time to mature and ferment. Patients with a genetic insufficiency of the diamine oxidase enzyme (DAO), and those that take medicine belonging to the group of monoamine oxidases (MAO), are particularly susceptible to the negative effects of amines. Diet plays an important role in the initiation, promotion, and progression of cancers. As a result of the illness, the consumption of some nutrients can be reduced, leading to nutritional deficiencies and resulting in malnutrition. Changes in metabolism may lead to cachexia in some patients suffering from NENs. The aim of this narrative review was to advance the knowledge in this area, and to determine possibilities related to dietary support. The authors also paid attention to role of biogenic amines in the treatment of patients with NENs. We can use this information to better understand nutritional issues faced by patients with gastroenteropancreatic neuroendocrine neoplasms (GEP-NENs), and to help inform the development of screening tools and clinical practice guidelines.

Keywords: gastroenteropancreatic neuroendocrine neoplasms (GEP-NENs); neuroendocrine neoplasms (NEN); neuroendocrine tumors; biogenic amines; nutrition; therapy

1. Introduction

Neuroendocrine neoplasms (NENs) are a heterogenic group of tumors originating from the endocrine glands (adrenal glands, pituitary gland, parathyroid glands), endocrine cells within gland tissues (pancreas, thyroid), or dispersed endocrine cells of the digestive and respiratory tracts [1]. The term neuroendocrine neoplasm is a general term which includes a group of well-differentiated neoplasms called neuroendocrine tumors (NETs), and a group of poorly differentiated forms called neuroendocrine cancers (NECs) [2].

The characteristic feature of some NENs is the ability to produce, store and secrete biogenic amines and peptide hormones, such as insulin, gastrin, vasoactive intestinal peptide, glucagon, or somatostatin [3–9].

Epidemiological NEN data show the incidence rate is about 5.6/100.000 per year [6]. NENs originating in the digestive system are called gastro-entero-pancreatic neuroendocrine neoplasms and represent 62–67% of NENs [7]. Small bowel NENs are the most common type of gastroenteropancreatic neuroendocrine neoplasms (GEP-NENs) originating from midgut and constituting about 38% of all GEP-NENs [10,11].

The occurrence of tumors within the small intestine, particularly the ileum, is estimated at about 2.8 to 8 cases per 1 million people each year. The incidence of neuroendocrine neoplasms of the duodenum is about 0.1 cases per 100,000 people, whereas, for the colon and the rectum, the values are respectively 7.8% and 13.7% of all NENs. The occurrence of stomach NEN does not exceed 2 cases per 100,000 people, whereas the value for the pancreas is about 4 to 12 cases per 1 million people per year [12].

NENs can originate everywhere in the body, not only in the gastrointestinal tract, pancreas, and lungs, but also in less frequent sites, such as the thymus, central nervous system, thyroid, skin, breast, and urogenital system [2]. The main location of NENs near the ileocecal valve often results in liver metastasis, which worsens the prognosis; the 10-year survivability decreases from 60% to 15–25%. The 5-year survivability of small intestine NETs is 73.8% in the case of local changes, but only 43.2% if distant metastasis occurs [12]. Moreover, malnutrition influences quality of life but also reduces tolerance to anti-cancer therapy and reduces survival in patients with cancer [13]. Currently, nutritional and vitamin status is a neglected area in patients with GEP-NENs [14]. Clinical practice guidelines and consensus guidelines for GEP-NENs with regards to best practice for diagnosis, treatment, and medical management are available, but the supportive care needs and optimal nutritional management of patients affected by these unique tumors remain under-researched [15]. The aim of this narrative review was to broaden and systematize knowledge in this area, determine the possibilities for dietary support and draw attention to the need for an anti-cancer diet rich in plant products and fiber. The aim was also to reduce along with the need to reduce at the same time for patients suffering from persistent diarrhea.

The authors paid attention to the role of alimentary biogenic amines in the genesis of general symptoms in the group of NENs patients, especially in those experiencing hypersensitivity to biogenic amines, and also focused on the different nutritional needs according to the severity of the disease and the patient's nutritional status.

We hope that our manuscript will contribute to the development of screening tools and clinical practice guidelines.

1.1. NEN Diagnostics

NEN diagnostics should include: biochemical measurements, imaging diagnostics and histopathological examination, which are crucial to diagnose and classify NEN.

For many years, the measurement of serotonin metabolites such as 5-hydroxyindoleacetic acid (5-HIAA) in 24 h urine collection was sufficient for most purposes, and has been considered the best method in laboratory diagnostics for functioning NENs with carcinoid syndrome [16]. Unfortunately, this method has limitations, and the consumption of pineapples, bananas, eggplant, the common walnut, paracetamol, caffeine, and naproxen can lead to false positive results, while, on the other hand, in patients treated with acetylsalicylic acid, adrenocorticotropin, levodopa and phenothiazine derivatives, false negative results can occur. There is a possibility of including the serum, platelet and urine serotonin concentrations as well.

A significant NEN diagnostic method is measuring chromogranins, particularly chromogranins A and B, which are proteins created and secreted by neuroendocrine tissues, fulfilling the role of non-specific NEN markers [17]. Chromogranin A is associated with the size of the tumor and allows for a better evaluation (in comparison to 5-HIAA) of recurrence in patients with a diagnosed NEN. It has been demonstrated that CgA levels in the plasma correlate with the load of the tumor and predict the survivability of patients with small intestine NENs. However, there is no correlation between the CgA plasma and the weight of the tumor, or the survivability of colorectal NENs [18], whereas high mRNA-binding protein 3 (IMP3) expression levels were determined to be associated with a high disease stage in patients with GEP-NENs [19].

In imaging diagnostics, somatostatin receptor scintigraphy (SRS) can still be considered [20]. The main factors that support this method include the fact that the dominating subtypes of the receptor on the GEP-NEN cells, SSR2 (somatostatin receptor type 2) and SSR5 (somatostatin receptor type 5), are the main point of uptake of the activity of octreotide (somatostatin analogue), and the presence of somatostatin receptors on the surface of 80–90% of GEP-NEN cells. SRS provides information on the location of the tumor as well as the degree of its development, and it allows the response to treatment with somatostatin analogues to be determined [20–22].

Because of greater diagnostic accuracy and lower radiation dose in an increasing number of centers, PET/CT with 68Ga-labeled somatostatin analogues has replaced SRS [22]. Other methods, such as ultrasound, colonoscopy, gastroscopy, endosonography, computed tomography (CT), magnetic resonance imaging (MRI), and various kinds of positron emission tomography (PET/CT) like F-FDG_PET/CT or the before mentioned 68Ga-DOTA-Peptide PET/CT, are available and useful for the localization of NENs, as well as for staging [23–25].

According to World Health Organization (WHO) 2017, based on WHO 2010 and European Neuroendocrine Tumor Society (ENETS) classifications that are crucial for the diagnosis of NENs, the histopathological examination should be supplemented with immunohistochemistry, which is based on Ki67 expression and allows for the division of NENs into three main groups of well differentiated tumors: NET G1 (low grade) with Ki67 $\leq$ 2%, NET G2 with Ki67 3–20% (intermediate grade), and NET G3 with Ki67 > 20% (high grade). Additionally, analysis of the morphology of the NEN cells is essential, in separate wells, for poorly differentiated NENs. Poorly differentiated NENs with Ki67 > 20% are called neuroendocrine cancers (NECs), among which we distinguish large-cell NECs (LC-NECs) and small-cell NECs (SL-NECs), usually with Ki67 > 55%. These classifications are not valid for extra GEP-NENs [25–28].

1.2. *Types of Hormonally Active Neuroendocrine Neoplasms (NENs)*

The clinical symptoms of neuroendocrine neoplasms of the digestive tract depend on the location of the tumor's primary site, and the amount of secreted peptide hormones and biogenic amines [8]. The most frequent initial symptom in patients with small intestinal NENs (siNENs) is abdominal pain resembling irritable bowel syndrome, but the great majority of GEP-NEN patients present symptoms characteristic of advanced cancer, such as anorexia, weight loss and fatigue; less than 5% of NETs are connected with a hormonal syndrome [24–27]. In hormonally active tumors, symptoms include hypoglycemic syndrome, carcinoid syndrome, Zollinger-Ellison syndrome, watery diarrhea-hypokalemia-achlorhydria syndrome (WDHA), and glucagonoma [28].

Presented below are classifications of NENs made on the basis of secreted substances; this is valid only for functioning forms that represent a minority of NENs.

A carcinoid is a hormonally active tumor originating from the central section of the digestive tract, characterized by the possible release of serotonin and other biologically active substances (kinin, tachykinin, dopamine, histamine, and prostaglandin) into the system's circulation, causing symptoms characteristic of carcinoid syndrome with accompanying liver metastases [29,30]. Carcinoid syndrome affects 4–10% of patients with small intestine NET location. The symptoms are present when liver metastasis occurs, and the produced serotonin, un-metabolized by hepatocytes, permeates directly into circulation. According to ENETS Consensus Guidelines Update 2016, carcinoid syndrome could be present in 20–30% of patients with metastases [26,31]. The slow growth of the tumor contributes to delayed diagnosis due to an asymptomatic course for many years. Carcinoid syndrome may manifest through explosive and watery diarrhea, present up to 30 times a day, which occurs for about 80% of carcinoid patients. The second most common symptom is paroxysmal skin flushes, from salmon-colored up to dark red, which affects about 85% of patients and usually occurs in the upper parts of the body (face, neck or chest). Common triggers include tyramine-containing foods (bananas, chocolate, blue cheese, red wine), alcohol, and stress. Carcinoid syndrome can be related to other symptoms: stomach aches, dizziness, telangiectasia, pellagra, tiredness, and sometimes the impairment of cognitive functions [32]. Hedinger syndrome occurs in over half of patients with carcinoid syndrome and can be the main cause of death due to right-side heart failure because of morphological changes and mechanical damage to the right heart valve apparatus due to un-metabolized serotonin [33,34]. For this reason, every patient with carcinoid syndrome requires cardiology consultation with echocardiography [26].

Insulinoma, originating from β cells, is the most frequent hormonally active pancreas NET, overproducing insulin and leading to hypoglycemia and hypoglycemic syndrome. The symptoms of hypoglycemia occur suddenly and paroxysmally in the morning due to significant fasting and after intense physical activity; they are accompanied by sweating, paleness, restlessness, shivering, palpitations, and hypersalivation [35]. The next stage of hypoglycemia includes psychomotor and concentration disorders, resulting in the loss of consciousness. Typical for insulinoma is a significant gain of body mass, mainly due to strong hunger caused by hypoglycemia, resulting in excessive caloric intake [34,35].

Gastrinoma is a gastrin-producing NET usually located in the pancreas or the front wall of the duodenum. In about 60% of patients, the tumor is malignant, resulting in metastasis, most often to the nearby lymph nodes and the liver [36,37]. Gastrin overproduction leads to overgrowth of parietal cells and increased secretion of stomach acid, causing stomach and duodenum ulceration. Furthermore, it deactivates pancreas enzymes, resulting in incorrect fat absorption and causing diarrhea. Severe ulcer disease combined with diarrhea in gastrinoma patients is known as Zollinger-Ellison syndrome (ZES). ZES patients typically suffer from stomach pain, vitamin B_{12} absorption disorders, a loss of body mass, colic, and kidney stones [36,37].

VIPoma is a NET producing vasoactive intestinal peptide (VIP). VIP is a neurohormone released by the central nervous system, intestines, pancreas, the respiratory tract, and the urogenital tract. VIP regulates the activity of smooth muscles, dilates blood vessels, and is responsible for water and electrolyte secretion by the digestive tract and inhibition of stomach acid secretion. Typically, the symptoms of excessive VIP secretion include watery diarrhea, hypoglycemia, achlorhydria, sometimes hypercalcemia or hypophosphatemia, and metabolic acidosis. The diarrhea volume usually exceeds 700 mL/24 h, and in 70% of cases can reach 3000 mL a day. Patients describe it as odorless, with a tea-like color. VIPoma can be connected with symptoms like lethargy, nausea, vomiting, and the weakening of muscles, as well as contractions that occur as a result of dehydration and hypoglycemia [34,38].

Glucagonoma originates from the α cells of pancreatic islets, which produce glucagon [39]. Its clinical image includes symptoms such as necrolytic migratory erythema (NME) (82%), usually located in the area of the lips and sexual organs, diabetes (80%), body mass loss (90%), a low level of zinc, niacin deficiency, abdominal pain, diarrhea, normochromic normocytic anemia (61%), and episodes of glossitis. Patients have thinning hair and dystrophic nails. Glucose intolerance in glucagonoma syndrome usually occurs proportionally to the size of the tumor. The concentration of glucagon in the plasma when fasting is higher in the group of patients with liver metastasis than in patients without accompanying metastasis. Liver metastasis reduces the ability of the liver to metabolize glucagon, increasing its concentration in peripheral blood [40].

Somatostatinoma originates from the cells of pancreatic islets which produce somatostatin. Somatostatin is an inhibitor of numerous secretory hormones, such as insulin, glucagon, gastrin, secretin, and motilin. Apart from strong inhibition, it has a direct influence on many target organs. It influences the activity of the intestines in terms of the absorption of nutrients, mainly fats and calcium. By stimulating prostaglandins, it slows down the secretion of stomach acid. The dominating illnesses in the clinical image of somatostatinoma are gallbladder stones, fat stools, body mass loss, and mild diabetes, which form somatostatinoma syndrome [34]. The most common location for the metastasis of this tumor is the liver, then lymph nodes and, lastly, the bones. The total removal of the tumor is usually very effective in the therapy of this illness [41,42]. It is worth mentioning that the presence of pancreatic neuroendocrine tumors (pNET) can occur as a part of inherited syndromes like multiple endocrine neoplasia type I (MEN 1), which could be responsible for 20–30% of gastrinomas, <5% of insulinomas, and rarely functional pNETs. Uncommon causes of pNETs include other inherited syndromes like von Hippel Lindau disease (VHL), neurofibromatosis type 1 and tuberous sclerosis [43].

2. The Treatment of Neuroendocrine Neoplasms

The treatment of choice is a surgical intervention with the curative or palliative aim, depending on the location and histopathology of the tumor. To qualify for surgery, patients have to be in generally good condition, with a tumor limited only to the primary site and the nearby lymph nodes. Patients with potentially resectable liver metastasis also qualify for the intervention [44,45]. Unfortunately, due to the presence of late clinical symptoms, NENs are usually diagnosed at advanced stages of the illness. This is why, in most cases, it is impossible to fully eliminate the changes [22].

Somatostatin analogues (SSAs) are an important part of NEN therapy and can be administered in neuroendocrine neoplasms (long-acting SSAs) as well as in a carcinoid crisis (short-acting SSAs). In treatment, we can use two types of SSA: octreotide and lanreotide. SSAs inhibit the secretion of many hormones, fulfill immunological, cytotoxic and cytostatic functions, and in specific conditions they can also be apoptotic through their direct influence on the somatostatin of tumor cells receptors (SSTR). In an indirect way, they lead to the inhibition of tumor mass factors, the proliferation of lymphocytes, and immunoglobulin synthesis. From an oncological point of view, the anti-proliferating effect of somatostatin analogues is the most important aspect, as it slows down the development of the illness and reduces the size of the tumor. It influences the digestive system in a multi-directional manner, slowing down the blood flow of visceral vessels as well as intestinal motility and transport. Most importantly, it inhibits the secretion of pancreatic and intestinal hormones. SSAs play a particularly important function in patients with hormonally active GEP-NENs. After they are administered, the symptoms associated with excessive secretion of biogenic substances are alleviated, improving the quality of life [44–47]. In the PROMID study, it was proven that long-acting repeatable octreotide acetate (octrotide LAR) significantly lengthens the time of tumor progression in patients with functionally active and inactive metastatic midgut NETs, and, a few years later, in the CLARINET study, it was shown that lanreotide therapy was associated with significantly prolonged progression-free survival (PFS) among patients with metastatic NETs of grade 1 or 2, with Ki-67 < 10% [48,49]. In the TELECAST study, it was proven that in patients with carcinoid syndrome not adequately controlled by SSA therapy, telotristat etiprate, an inhibitor of tryptophan hydroxylase, can be used to limit the synthesis of serotonin [50].

In patients in an advanced stage of the disease, as well as in those with relapses after primary therapy and who did not undergo full surgical treatment, other types of therapy, like peptide receptor radionuclide therapy (PRRT), Tyrosine Kinase Inhibitor (TKI), mTOR inhibitors, or chemotherapy, can still be used [24].

PRRT is another type of GEP-NEN treatment based on "a combination of somatostatin analogues with yttrium or lutetium isotopes, and the cytotoxic factor is the ionizing radiation of the isotope" [51]. Radionuclide therapy seems to be a good method when patients intensely accumulate the marker at each neoplasm site of a small size, which can be used to achieve total remission, or at least a reduction in the neoplasm's mass [52]. In the presence of liver metastases by GEP-NEN, as a form of palliative treatment, radiofrequency ablation (RFA), trans-arterial embolization (TAE), and trans-arterial chemoembolization (TACE) can be offered [53].

Multi-target tyrosine kinase inhibitors (MTKIs), such as axitinib, cabozantinib, famitinib, lenvatinib, nintedanib, pazopanib, sorafenib and sulfatinib, represent a new approach to NEN treatment [54]. Sunitinib malate has been approved by regulatory agencies for pancreatic NENs [55]. Sunitynib is an oral multi-targeted inhibitor of various receptor tyrosine kinases that leads to a decrease in angiogenesis, growth, proliferation, and metastatic spread [56].

According to the RADIANT-3 and RADIANT-4 studies, the mTOR inhibitor ewerolimus has an established place in the therapy of advanced and progressive pancreatic NETs and non-functional lung and gastrointestinal NETs [57,58].

When metastasis is present or in case of disease progression in patients with NETs chemotherapy can be used. Capacitabine and Temozolemide (CAPTEM) shows significant activity in patients with metastatic well-differentiated pancreatic NETs [59]. At the same time, CAPTEM presents significant activity in patients with metastatic grades 2 and 3 pancreatic and non-pancreatic NETs with manageable toxicity. Systemic combined chemotherapy like cisplatin + etoposide, streptozocine + 5-fluorouracil, streptozocin + doxorubicin, leucovorin + 5-fluorouracil + oxaliplatin (FOLFOX) or leucovorin + 5-flurouracil + irinotecan (FOLFIRI) has been designed to treat NEC patients according to primary tumor localization [60,61].

In the randomized clinical trials (RCT) of neuroendocrine tumors, 22 different therapy strategies were compared, stating that there are a number of effective therapies with different safety profiles available to patients, suggesting an overall superiority of combination therapies [62]. For patients with advanced NETs as a best second- and third-line treatment, respectively, to progression-free survival (PFS), PRRT, SSA + bevacizumab, and SSA + interferonalfa should be considered [63].

In the absence of an optimal treatment strategy, other methods should be found. There are several future potential NEN therapies, such as immunotherapy (programmed death ligand 1, cytotoxic T-lymphocyte antigen-4 blockers) and somatostatin-dopamine multi-receptor chimeras [64].

3. The State and Method of Nutrition with Reference to the Risk of Cancer

Epidemiological studies strongly suggest that BMI and especially visceral fat accumulation, decreased physical activity, and unhealthy diets are key elements in the pathogenesis and prognosis of many common cancers. The phenomenon known as "the obesity paradox" suggests a potentially protective effect in patients with overweight or slight obesity, increasing their survivability after diagnosing a neoplastic disease. [65].

The accumulation of multiple DNA mutations in critical genes (oncogenes or tumor suppressor genes) of particular cells, if not properly controlled through the induction of senescence or apoptosis, can lead to uncontrolled cell proliferation and the progressive transformation of cells into highly malignant tumor cells [66]. Calorie restrictions without malnutrition are the most potent and reproducible physiological intervention for increasing lifespan and protecting against cancer [67].

Not many nutrients have a cause and effect relation with cancer, some of them include: fried, smoked or roasted red meat, food contaminated with aflatoxin, preserved salty meals, excessive alcohol consumption [67,68]. The risk of cancer can also be reduced by introducing a diet rich in plant food (e.g., vegetables, beans, fruit and wholegrain products) and by limiting the consumption of animal fat, meat and fatty dairy products [68].

One such diet is the Mediterranean diet (MD), which could influence the reduction of the aggressiveness of different tumor types and tumor size [16,69]. The diet plays an important role in the initiation, promotion, and progression of cancers [70]. Vegetables and fruits are important sources of a wide variety of micronutrients and other bioactive compounds, including antioxidants, vitamins, folates, carotenoids, glucosinolates, indoles, isothiocyanates, protease inhibitors, and phytochemicals, such as lycopene, phenolic compounds, and flavonoids, which have been demonstrated to exhibit anticancer properties [70,71]. All these compounds may act against cancer through different mechanisms, including their antioxidant, anti-mutagenic, and anti-proliferative properties. Furthermore, there is a connection between obesity and the increased risk of endometrium, breast, colon, esophagus, kidney, pancreas, gallbladder or liver cancers [72]. Obesity and an excess of fat tissue resulting from a chronic energy imbalance are associated with the increased risk of oxidative stress. Consequently, there are disorders of lipid and carbohydrate metabolism, insulin resistance, systemic inflammation, changes in hormone levels and growth factor concentrations which have a key role in the pathogenesis of many neoplasms.

3.1. Dietary Recommendations for NEN Patients

Nutrition care plans are an integral part of the multidisciplinary management of patients with NETs. Nutritionists with expertise in NETs can provide dietary approaches to improve the quality of life and nutritional status during therapeutic modalities used for patients with NETs and particular in palliative care [73]. Unfortunately, there are not enough registered physicians and dieticians who have expertise in the nutritional management of NETs [74]. Factors such as unhealthy diets, tobacco, alcoholism, infections and occupational exposures contribute to the formation of neoplasms [75]. The need for consistent dietary guidelines for NEN patients and collaboration with nutritionists in multidisciplinary healthcare teams in NET management have been emphasized by other authors [69–76]. Recommendations for a healthy diet are based on the 2015–2020 Dietary Guidelines Advisory Committee for patients with newly diagnosed asymptomatic NETs [77]. Therefore, a diet with five servings of vegetables and fruits, including legumes and with meat restrictions, will be the best solution for newly diagnosed and asymptomatic patients [78,79]. However, the diet will depend on the symptoms of each patient, the stage of the disease, the type of therapeutic management, and the individual's nutritional status. The best diet for a NEN patient is, therefore, an individualized diet. Moroever, dietary recommendations in neuroendocrine neoplasms should take into account the excessive production of hormones, the source of which could be endogenous or exogenous. Furthermore, an excess of biogenic amines in the body often leads to the occurrence of specific symptoms, such as diarrhea, nausea, vomiting, and metabolic disorders in the form of hypoglycemia and hyperglycemia, contributing to malnutrition and the general weakening of the patient [80–82].

3.2. The Causes of Diarrhea in NEN Patients and Dietetic Modifications

The presence of diarrhea in NEN patients may result from various factors, including the metabolism of hormonally active NEN, biogenic amines, drugs and niacin deficiency.

Diarrhea occurs independently of the consumed meal. It can occur after the meal, especially after the consumption of large amounts and fatty products, but it can also occur after fasting and at night. In the case of some neoplasms (somatostatinoma and, less often, gastrinoma), diarrhea is associated with impaired digestion or the absorption of fatty acids in the digestive tract. Lastly, diarrhea can be caused by surgical intervention in the area of the intestine, which leads to a direct loss of absorbency [41,42,83].

Biogenic amines are nitrogen compounds that are present in products in two forms: naturally, due to synthesis by plants, animals and microorganisms, or as an additive during production, in the form of preservatives [84]. Their synthesis occurs as a result of the breakdown of peptides and proteins. They then undergo further transformations, creating new, different amines. Physiologically, they participate in numerous processes; they are a source of nitrogen and they are precursors of the synthesis of hormones, nucleic acids, and proteins [85]. The synthesis of amines in food depends on the availability of appropriate amino acids and bacteria, as well as on the environmental conditions, which determine the correct activity of enzymes and bacterial growth. Their activity is observed in numerous food products, including drinks. Figure 1 shows products that include significant amounts of amines, as well as products that should be eliminated from the diet with some GEP-NENs [86,87].

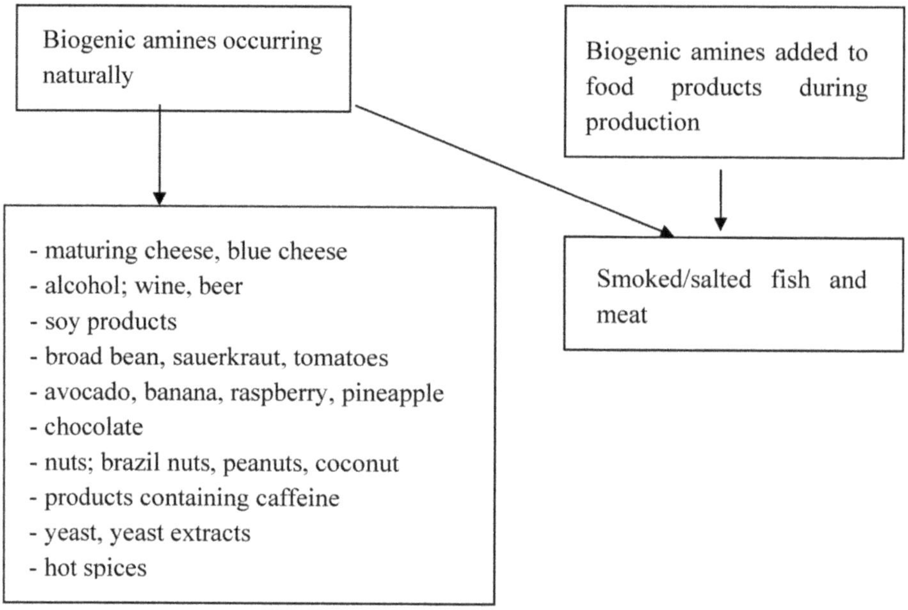

Figure 1. Presence of amines in food products.

The most frequent causes of symptoms are histamine, tyramine and putrescine. Patients with a genetic deficiency in diamine oxidase (DAO), as well as those that take medicine belonging to the group of monoamine oxidases, are particularly susceptible to the negative effects of amines [85,86]. Histamine is formed from histidine with the co-participation of histidine decarboxylase enzyme [87]. Mastocytes, basophils, and enterochromatophilic cells of the digestive tract are the most significant cells that produce endogenic histamine. They have the ability to store this amine in cytoplasmic granulation and, in the subsequent stage, to release the amine into blood circulation through immunological and non-immunological stimuli. Histamine metabolism proceeds in two ways [88]. The first features methylation into N-methylhistamine via N-methyltransferase (HNMT). Thanks to this enzyme, histamine can be metabolized only intracellularly. The second method is oxidation into imidazoleacetic acid thanks to the activity of diamine oxidase (DAO). Disease of the digestive tract, as well as the activity of alcohol and some medicines, can be the cause of a secondary shortage of DAO. The role of histamine is mainly based on the extracellular metabolism of the described biogenic amine, so it is reduced after the consumption of products that are a rich source of histamine. Exogenic sources of histamine are products that take a long time to mature and ferment, such as baker's yeast, red and white wine, beer, champagne, kefir, blue cheese, cheese spread, yellow cheese, prosciutto, salami, highly processed cold meat, smoked fish, avocado, spinach, eggplant, sauerkraut, ketchup, and various spices and herbs [83,84,89,90]. Tyramine can be the cause of reddening, which is one of the most common clinical carcinoid symptoms. It is an aromatic monoamine whose systematic name is 4-hydroxy-phenethylamine. Numerous studies have demonstrated that tyramine is the most common biogenic amine in cheese. It is found in the highest content in veined blue cheese, such as gorgonzola or roquefort. The amount varies in specific parts of the cheese, with the highest content in the external part. The primary producers of tyramine in cheeses are Gram-positive bacteria, such as Lactobacillus, Enterococcus, Leuconostoc, Lactococcus and Carnobacterium [88,89]. Moreover, it is also common in fermented foods, such as soy sauce, shrimp spread, marmite, eggplant, spinach, sauerkraut, sausages, ham, smoked fish, anchovies, sardines, beer, wine, and chocolate. The characteristic term associated with tyramine poisoning is the "cheese effect", which is highlighted by a hypertensive crisis in blood pressure. There have been numerous cases of death resulting from myocardial infarction or stroke as a consequence of consuming food products with significant amounts of tyramine [89].

Out of all of the symptoms of GEP-NENs, the most common are skin reddening and diarrhea, the latter of which leads to dehydration and electrolyte disturbances [2,5,90,91]. They appear as a result of the promotion via the secretion of hormones, peptides, and amines, and the excessive secretion of liquids to the intestinal mucosa [91,92]. Dietary treatment recommended avoiding hot, seasoned, fatty and overly large meals when suffering with diarrhea. One should consume meals that include proteins, mainly lean products of meat, poultry, curd cheese, eggs, and yoghurt. As a source of carbohydrates, starch products such as rice and finely ground oats are recommended. Simple sugars (glucose and fructose) should be eliminated from the diet because they strengthen fermentation processes. It is recommended to exclude lactose and saccharose for a few days. The following products are rich in the aforementioned sugars: jams, honey, candy, and apple and grape juice (which should be particularly avoided). Vegetables should be mild. They should not cause excessive production of flatulence and are best served shredded and boiled. The best are those with high amounts of pectin, such as pumpkin and carrots, and when it comes to fruit, the best ones are apples and bananas. Meals should be served with plant fats to strengthen the energy properties of the diet and reduce the glycemic index. It is also crucial to supply liquids (potassium water in particular) to avoid dehydration. It is recommended to consume two liters per day [93,94]. One should eliminate liquids that include caffeine and strong tea because they do not hydrate well enough and they are rich in biogenic amines [95,96]. There are only a few references to the use of enteral nutrition in the clinical guidelines of patient management with oncology treatment-related diarrhea. Although no data is available for patients with NENs, it appears that the inclusion of oligomeric enteral nutrition formula in patients with diarrhea and malnutrition may be justified [97]. The essential element will be to determine the functional capacity of the patient's intestine and nutritional status. Therefore, nutrition in this case may be based on typical dietary recommendations, through supplementation with oral oligomeric enteral nutrition [98] along with full enteral nutrition with oligomeric formula, up to potentially complete parenteral nutrition [99]. When it comes to the share of dietary supplements, products with high osmolality should be avoided [100], and supplements could even be included in cases of severe diarrhea or severe malnutrition [101]. It is important to be cautious when using nutraceuticals or other dietary supplements because of the possibility of disrupting chemotherapy. However, due to the deficits observed in these patients, it is important to conduct chemotherapy [66,102,103].

3.3. *Procedures to Follow in the Case of Constipation in NEN Patients*

In some cases, constipation can appear in NEN patients. It is usually accompanied by abdominal distension and flatulence [104]. The main cause is the ileus, or a side effect of the applied pharmacological therapy. The excessive secretion of catecholamines by the tumor can reduce the peristaltic activity of the digestive tract, leading to chronic constipation [34]. The undertaken nutritional intervention should cover the increased supply of liquids in the form of mineral water, juice, and chamomile or fennel tea. It is recommended to consume products with high amounts of insoluble dietary fiber such as Graham bread, bran bread, thick groats, or oatmeal. Inulin in particular has properties that support the struggle with constipation [105]. It has also been found that the low fermentable oligosaccharides, disaccharides, monosaccharides, and polyols (FODMAP) formula improves diarrhea and nutritional status in hospitalized patients [106]. Cruciferous and bulbous vegetables are not recommended due to causing excessive flatulence [85]. Moreover, in patients with chronic constipation, dysbiosis is observed in the area of the small intestine, with Clostridium and Enterobacteriaceae as the dominating bacteria. This is why it is so important to include probiotic bacteria that produce short-chain fatty acids (*Lactobacillus* and *Bifidobacterium*) in order to reduce the pH of the intestine, to stimulate intestinal motility and to accelerate the transit of stool [106,107]. Regular physical activity is also beneficial [65,66]. For a better understanding of the individual nutritional approach to the patient with NEN, considerations are presented in Table 1A.

3.4. Nutrition that Takes into Account the Hormone Activity of NEN

When analyzing all nutritional aspects, it is also important to consider the hormone activity of NEN, which has an influence on metabolic changes in patients.

As a result of the excessive production of insulin, there is a reduction in the level of glucose, which can, in serious cases, be the cause of death. The aim of dietetic therapy is to prevent long fasting between meals via the frequent consumption of small portions of food during the day and at night. A high-protein diet is also important as glucose can be metabolized by the organism for a longer period of time, and its secretion to the circulation is slower. Thanks to that, the risk of secondary hypoglycemia decreases, and the increase in body mass, which is characteristic of insulinoma, is unnoticeable. Products with a low glycemic index and complex carbohydrates maintain the level of glucose at a stable level, preventing post-meal hypoglycemia. When the level of glucose in the blood decreases, it is important to immediately supply carbohydrates with a high glycemic index like fruit juice, because they are quickly absorbed [39–41].

The reverse effect, hyperglycemia, usually develops secondarily, most often in patients predisposed to diabetes. In the case of NEN, hyperglycemia develops due to the direct influence of the tumor mass on the pancreas, reducing the amount of insulin, as a result of surgical treatment, and also due to pharmacological treatment with a somatostatin analogue (SSA). This effect of treatment with a somatostatin analogue may be caused by impaired fasting glucose (IFG), impaired glucose tolerance (IGT), and diabetes mellitus (DM). These patients should have a healthy, balanced diet, based on products with a low glycemic index. Food products should be combined with fats, which will help keep glucose levels constant. High consumption of fiber should also be considered [15,96].

Another issue that has to be taken into account is the common niacin deficiency and risk of pellagra, which occurs in patients with carcinoid syndrome [108]. Vitamin B_3 deficiency results from the increase of the metabolism of tryptophan into serotonin (Figure 2).

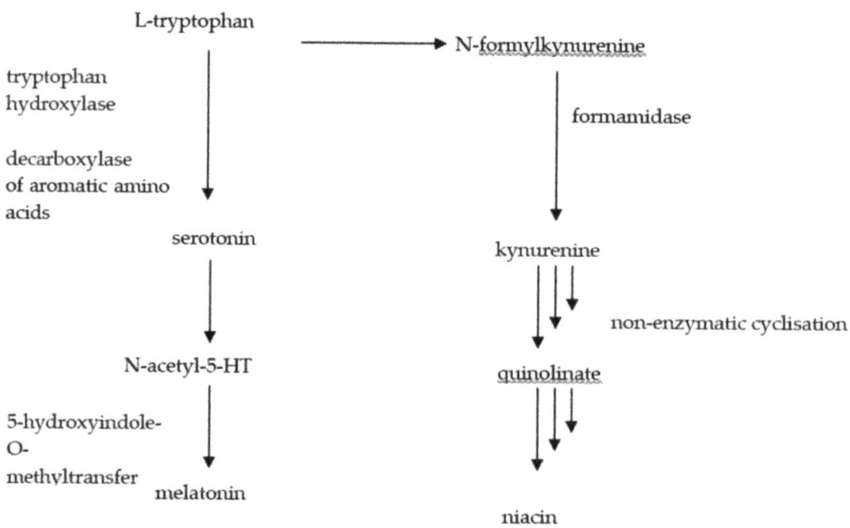

Figure 2. Transformation of tryptophan.

The deficiency leads to skin inflammation, diarrhea and mental disturbance, which can lead to death in severe cases if left untreated [109]. The skin becomes rough, fractures easily, and discoloring occurs. The most noticeable changes are to the face, neck and hands. Patients should be supplied with food rich in this vitamin, such as liver, fish, meat, yeast, wheat bran and the seeds of leguminous plants. In smaller amounts, niacin is also found in fruit and vegetables, bakery products, and milk [57].

When the patient is unable to cover their needs through consumed food, supplementation of this vitamin should be taken into consideration, supplying from 25 to 50 mg/day [16].

The disorders of digestion and/or the absorption of fatty acids that cause fatty diarrhea may be the result of gastrinoma or—less frequently—somatostatinoma. This is why it is important to reduce the supply of fats in the diets or to propose a substitution of pancreatic enzymes [110]. For a better understanding of the individual nutritional approach to the patient with NEN, considerations are presented in Table 1B.

3.5. A Diet for NEN Patients under the Risk of Malnutrition and/or Cachexia

Two studies have indicated that as many as one in four NET patients are malnourished, as assessed using the Malnutrition Universal Screening Tool and Subjective Global Assessment (SGA) tool [111]. Screening for malnutrition should be a part of routine care in every GEP-NEN patient [98]. Malnutrition has substantial negative consequences for cancer patients including increased mortality, poorer quality of life, and increased health care costs [112]. Malnutrition, which leads to the devastation of the body, is often associated with cancer cachexia, which is characterized by a loss of fat and muscle mass. The main cause of malnutrition is limited food consumption by the patient, which might amplify the symptoms associated with the treatment, such as nausea, vomiting, inflammation of the mucous membrane, abnormal absorption, anorexia, tiredness, and pain [113]. The reasons for the development of cachexia are (among other) metabolites produced by the tumor, which can cause anorectic effects in the center of hunger and satiety located in the brain. Another factor is systemic inflammation, which amplifies hypermetabolism, body mass loss, and tiredness. There are several studies reporting malnutrition in NENs. The range of reported malnutrition is 4.9%–38% in the course of progressive disease [114,115]. Omega-3 fatty acids have a positive effect on the treatment of neoplastic cachexia [116]. To cover the increased caloric needs, the patient's diet should include an appropriate amount of all nutrients: proteins, fat, carbohydrates, vitamins and minerals. Sometimes, apart from meals, it will also be important to supply patients with dietary supplements that include necessary nutrients [95]. Arginine regulates the production of NO in cancer and thus in might support the development of anti-cancer drugs that target this key metabolic pathway [117]. The diet should prevent body mass loss, lead to the reconstruction of tissues, and improve the way the patient feels. It is recommended they consume meals more frequently, but in smaller amounts. It is also recommended patients eat snacks between meals. It is not recommended to drink between meals due to the excessive dilution of gastric juice, which disturbs digestive processes and increases food volume. In patients that have problems with the consumption of meals or in severely malnourished patients, it is recommended to introduce enteral nutrition or parenteral feeding [7]. The content of the applied mixture should be individually adjusted to the needs of the patient, taking into consideration his illnesses and providing all the necessary nutrients [118].

4. Dietary Care Taking into Account Pharmacotherapy

It is important to consider the interaction of some medicines with food and the changes in the secretion by some organs, such as the pancreas. During the investigation it was determined that the treatment with everolimus or sunitinib may pose a risk for patients because there is an interaction with food (grapefruit, camomile, cranberry, garlic, ginseng, green tea extract, pepper, resveratrol and soy) associated with the inhibition of P450 (CYP) 3A pathway, which may lead to the toxicity of the medicine. Moreover, high-fat meals are inhibitors of tyrosine kinase [66]. Furthermore, temozolomide should not be supplied with food because fat—by changing the pH of the stomach—inhibits CYP P450. It has also been demonstrated that somatostatin analogues that are common in the treatment of advanced well-differentiated NEN may lead to exocrine pancreatic insufficiency in some patients [119]

For a better understanding of the individual nutritional approach to the patient with NEN, considerations are presented in Table 1C.

Table 1. (**A**) Proposed dietary care solutions for patients with Nuroendocrine neoplasm (NEN) according to the patient's nutritional status (BMI). (**B**) Proposed solutions for the dietary care of patients with NEN, taking into account NEN hormone activity. (**C**) Sample pharmacotherapy of NEN patients taking into account interactions with food.

(**A**) Proposed dietary care solutions for patients with NEN according to the patient's nutritional status (BMI).		
Nutritional status (BMI)	Symptoms	Dietary care solutions
>30 Obesity, visceral fat accumulation	No persistent bothersome symptoms	Anti-neoplastic diet (based on the high quantity and diversity of plant products) or Mediterranean and additionally reduction diet [69–71,74–79] Regular physical activity adjusted to the patient's capabilities [77–79].
	Severe diarrhea with progressing reduction of body mass	Consider supplementation, especially with omega 3 [94,116]; diet that includes a reduced amount of fiber, legumes and brassicas until symptoms decline [15,80,94]; Elimination of biogenic amines [87,88,92]; Electrolyte supplementation [91]; Potassium supplementation (hypokalemia) [15]; Avoid hot, very spicy, fatty and large meal portions [66]; Consume meals that include proteins, mainly lean poultry, cottage cheese, eggs and yoghurt; Carbohydrates: rice and finely ground oats, pumpkin, carrot, bananas, apples [66]; Exclude lactose, saccharose, fructose and glucose [15,66,93].
	Constipation	Anti-neoplastic, Mediterranean diet [69–71,74–79] high in fiber with inulin [15,102], consider probiotics therapy: *Lactobacillus acidophilus* and *Bifidobacterium lactis* [106]; Increased supply of liquids (mineral water with lemon, aloe, additionally drank in the morning) [34]; Regular physical activity, physiotherapeutic massage [74–76].
	Disturbed carbohydrate metabolism	Low glycemic index diet with limited amounts of fruit (glucose, fructose, saccharose), supplemented with MUFA and PUFA [66–73,117]; Regular physical activity [77–79].
* 26–29.9 overweight Visceral Fat accumulation <26	No chronic, irritating symptoms	Anti-neoplastic, Mediterranean diet. Perhaps consider a reduction diet if the patient's diet did not decrease recently due to the intense course of the disease [69–71,74–79]; Regular physical activity [77–79].
	Irritating diarrhea with progressing reduction of body mass	Procedures the same as in the case of diarrhea >30 BMI.
	Constipation	Procedures the same as in the case of constipation >30 BMI.
	Disturbed carbohydrate metabolism	Procedures the same as in the case of disturbed carbohydrate metabolism >30 BMI.

Table 1. *Cont.*

(**A**) Proposed dietary care solutions for patients with NEN according to the patient's nutritional status (BMI).		
Nutritional status (BMI)	Symptoms	Dietary care solutions
26–22/23 **	No chronic, irritating symptoms	Anti-neoplastic, Mediterranean diet according to needs of the body [69–71,74–79] Regular physical activity [77–79].
	Irritating diarrhea with progressing reduction of body mass	Procedures the same as in the case of diarrhea >30 BMI.
	Constipation	Procedures the same as in the case of constipation >30 BMI.
	Disturbed carbohydrate metabolism	Procedures the same as in the case of disturbed carbohydrate metabolism >30 BMI.
<22/23 ** At the risk of malnutrition	No chronic, irritating symptoms	Anti-neoplastic, Mediterranean diet [69–71,74–79] Stimulation of tissue reconstruction, e.g., through the incorporation of industrial diet preparations that additionally feature arginine [117].
	Irritating diarrhea with progressing reduction of body mass	Incorporation of oligomeric formula of enteral nutrition in patients with diarrhea and progressing malnutrition [97,98]; Potentially–full parenteral nutrition [101]; Diets with reduced osmolarity [100]; Electrolyte supplementation [91]; Incorporate multi-element supplementation that includes omega-3 [95].
	Cachexia	Enteral nutrition and parenteral nutrition, omega-3 supplementation [102], multi-element supplementation [89,95].
(**B**) Proposed solutions for the dietary care of patients with NEN taking into account NEN hormone activity		
NEN	Symptoms	Nutrition
Carcinoid	Increased metabolism of tryptophan into serotonin/spastic diarrhea	Supplementation of niacin deficiency (vitamin PP), supplementation 25–50 mg/day [16]; Include the consumption of fish, meat, bran and the seeds of legumes [106,107]; Regular physical activity after the earlier analysis of the heart using echocardiography [72].
Gastrinoma	Increased gastric acid synthesis and inactivation of pancreatic enzymes. Disorders of digestion and/or absorption of fatty acids = fatty diarrhea	Consume meals that include fats, mainly lean poultry, cottage cheese, eggs and yoghurt; Carbohydrates: rice and finely ground oats, pumpkin, carrot, bananas, apples; Limit fats or include pancreas enzymes' substitution [88,89]; Regular physical activity [72].
Somatostatinoma	Inhibition of the exocrine pancreatic function/steatorrhea	Procedures the same as in the case of gastrinoma.

Table 1. *Cont.*

(**B**) Proposed solutions for the dietary care of patients with NEN taking into account NEN hormone activity		
NEN	Symptoms	Nutrition
Vipoma	Water and electrolyte secretion by the digestive tract and inhibition of stomach acid secretion/secretory diarrhea	Special care for hydration and electrolyte management [91].
Glucagonoma	Disturbed carbohydrate metabolism; glucagon overproduction; impaired fasting glucose/impaired glucose tolerance/diabetes	Low glycemic index diet with the limitation of fruit; Prevention of long fasts between meals during the night break; Regular physical activity [72].
Insulinoma	Disturbed carbohydrate metabolism insulin overproduction/hypoglicemia	In the case of frequent hypoglycemia in insulinoma, the supply of carbohydrates with a high glycemic index, e.g., fruit juice [34–41]; Prevention of long fasts between meals during the night break.
(**C**) Sample pharmacotherapy of NEN patients taking into account interactions with food		
Medicine	Influence	Food
Everolimus, sunitinib [66]	P450 (CYP) 3A4 inhibition	Exclude for the diet: grapefruit, camomile, cranberry, garlic, ginseng, green tea extract, pepper, resveratrol and soy
Sorafenib	Inhibitors of tyrosine kinase	High fat meals
Capecytabine	is unstable under strongly acidic conditions	should be administered with a meal (up to 30 min after a meal)
Temozolomide [66]	CYP P450 inhibition through stomach pH	Not to be supplied together with food (on empty stomach)
Long-acting somatostatin analogues [110,119]	Exocrine pancreatic insufficiency	Include the substitution of pancreatic enzymes

* higher survival rate [65]; ** women/men.

5. Summary

Neuroendocrine neoplasms of the digestive tract, through the secretion of hormones, peptides and biogenic amines, cause various symptoms that can be reduced to improve the quality of life through a balanced diet and physical activity. The inclusion of low glycemic index products in the diet prevents the abrupt decrease and fluctuations in the level of glucose in the blood. This illness may result in the limitation of the consumption of some nutrients, leading to the development of nutritional deficiencies, excessive body mass loss and, eventually, malnutrition. Cachexia can be observed in some NEN patients. It can occur as the result of the tumor's production of metabolites that have an anorectic effect on the center of hunger and satiety in the brain. This is why enteral nutrition support and parenteral feeding should be considered in these patients. The frequent presence of diarrhea can be amplified by the consumption of biogenic amines. Therefore, patients that suffer from this problem should eliminate the following products from their diet: smoked fish and meat, soy products, avocado, raspberries, pineapples, chocolate and nuts. The carcinoid syndrome is characterized by the deficiency of niacin that results from the increased metabolism of tryptophan into serotonin, which can explain frequent changes in moods. A well-balanced nutritional plan not only supports the struggle against the illness, but it also eliminates the side effects of therapies, improving the quality of life.

5.1. *Structure of the Underlying Research*

The present narrative review evaluates the above-mentioned topics by considering the literature published up to 31 December 2019. A literature search was conducted utilizing the PubMed and Scopus databases. The terms used were: neuroendocrine tumors, neuroendocrine neoplasms, gastroenteropancreatic neuroendocrine neoplasias, and biogenic amines. The passwords were checked on terms: neuroendocrine tumors, neuroendocrine neoplasms, gastroenteropancreatic neuroendocrine neoplasias, and biogenic amines. These terms were combined with diagnosis, treatment, nutrition, diarrhea, constipation, and nutrition assessment. Studies that were not in English, letters to editor, and abstracts to conferences were excluded. All included studies were screened and discussed by the authors until a general consensus was reached.

Ethical Approval

This article does not contain any studies with human participants or animals performed by any of the authors.

Author Contributions: Conceptualization, J.P, A.J, M.S, Search strategy and literature search: J.P., M.S Supervision and mentorship, M.S., A.S. Evaluation of study quality and bias: J.P., A.J., A.S., E.A.-M., M.S., methodology, M.S., J.P. writing–original draft preparation, J.P., A.J, M.S, writing–review and editing M.S., J.P., Funding: J.P., A.S., E.A.-M. All authors have read and agreed to the published version of the manuscript

Funding: The research received no external funding.The APC was funded by Fundacja Na Rzecz Rozwoju Medycyny w Dziedzinie Endokrynologii i Chorob Metabolicznych, NIP 9552334867, REGON 321229151, ul. Z. Nalkowskiej 4, 70-785 SZCZECIN, Poland.

Conflicts of Interest: The authors declare no conflict of interest.

References

1. Schimmack, S.; Svejda, B.; Lawrence, B.; Kidd, M.; Modlin, I. The diversity and commonalities of gastroenteropancreatic neuroendocrine tumors. *Langenbecks Arch. Chir.* **2011**, *396*, 273–298. [CrossRef] [PubMed]
2. Oronsky, B.; Ma, P.C.; Morgensztern, D.; Carter, C.A. Nothing but NET: A Review of Neuroendocrine Tumors and Carcinomas. *Neoplasia* **2017**, *19*, 991–1002. [CrossRef] [PubMed]
3. Scherübl, H.; Cadiot, G. Early Gastroenteropancreatic Neuroendocrine Tumors: Endoscopic Therapy and Surveillance. *Visc. Med.* **2017**, *33*, 332–338. [CrossRef] [PubMed]
4. Rindi, G.; Wiedenmann, B. Neuroendocrine neoplasms of the gut and pancreas: New insights. *Nat. Rev. Endocrinol.* **2011**, *8*, 54–64. [CrossRef]

5. Öberg, K.; Hellman, P.; Kwekkeboom, D.; Jelic, S. On behalf of the ESMO Guidelines Working Group Neuroendocrine bronchial and thymic tumours: ESMO Clinical Practice Guidelines for diagnosis, treatment and follow-up. *Ann. Oncol.* **2010**, *21*, v220–v222. [CrossRef] [PubMed]
6. Klöppel, G. Neuroendocrine Neoplasms: Dichotomy, Origin and Classifications. *Visc. Med.* **2017**, *33*, 324–330. [CrossRef]
7. Taal, B.; Visser, O. Epidemiology of Neuroendocrine Tumours. *Neuroendocrinology* **2004**, *80*, 3–7. [CrossRef]
8. Meeker, A.K.; Heaphy, C. Gastroenteropancreatic endocrine tumors. *Mol. Cell. Endocrinol.* **2014**, *386*, 101–120. [CrossRef]
9. Srirajaskanthan, R.; Ahmed, A.; Prachialias, A.; Srinivasan, P.; Heaton, N.; Jervis, N.; Quaglia, A.; Vivian, G.; Ramage, J. ENETS TNM Staging Predicts Prognosis in Small Bowel Neuroendocrine Tumours. *ISRN Oncol.* **2013**, *2013*, 420795. [CrossRef]
10. Cheung, V.T.F.; Khan, M.S. A guide to midgut neuroendocrine tumours (NETs) and carcinoid syndrome. *Front. Gastroenterol.* **2014**, *6*, 264–269. [CrossRef]
11. Jernman, J.; Valimaki, M.J.; Louhimo, J.; Haglund, C.; Arola, J. The Novel WHO 2010 Classification for Gastrointestinal Neuroendocrine Tumours Correlates Well with the Metastatic Potential of Rectal Neuroendocrine Tumours. *Neuroendocrinology* **2012**, *95*, 317–324. [CrossRef] [PubMed]
12. Modlin, I.; Champaneria, M.C.; Chan, A.K.; Kidd, M. A Three-Decade Analysis of 3,911 Small Intestinal Neuroendocrine Tumors: The Rapid Pace of No Progress. *Am. J. Gastroenterol.* **2007**, *102*, 1464–1473. [CrossRef] [PubMed]
13. Arends, J.; Bachmann, P.; Baracos, V.; Barthelemy, N.; Bertz, H.; Bozzetti, F.; Fearon, K.; Hütterer, E.; Isenring, E.; Kaasa, S.; et al. ESPEN guidelines on nutrition in cancer patients. *Clin. Nutr.* **2017**, *36*, 11–48. [CrossRef] [PubMed]
14. Fearon, K.C.H.; Strasser, F.; Anker, S.; Bosaeus, I.; Bruera, E.; Fainsinger, R.L.; Jatoi, A.; Loprinzi, C.; Macdonald, N.; Mantovani, G.; et al. Definition and classification of cancer cachexia: An international consensus. *Lancet Oncol.* **2011**, *12*, 489–495. [CrossRef]
15. Gallo, M.; Muscogiuri, G.; Pizza, G.; Ruggeri, R.M.; Barrea, L.; Faggiano, A.; Colao, A. On behalf of NIKE Group The management of neuroendocrine tumours: A nutritional viewpoint. *Crit. Rev. Food Sci. Nutr.* **2017**, *59*, 1046–1057. [CrossRef]
16. Lindström, M.; Tohmola, N.; Renkonen, R.; Hamalainen, E.; Schalin-Jäntti, C.; Itkonen, O. Comparison of serum serotonin and serum 5-HIAA LC-MS/MS assays in the diagnosis of serotonin producing neuroendocrine neoplasms: A pilot study. *Clin. Chim. Acta* **2018**, *482*, 78–83. [CrossRef]
17. Tritschler, S.; Erdelkamp, R.; Stief, C.; Hentrich, M. Neuroendokrines Prostatakarzinom. *Der Urol.* **2017**, *56*, 1475–1484. [CrossRef]
18. Koenig, A.; Krug, S.; Mueller, D.; Barth, P.J.; Koenig, U.; Scharf, M.; Ellenrieder, V.; Michl, P.; Moll, R.; Homayunfar, K.; et al. Clinicopathological hallmarks and biomarkers of colorectal neuroendocrine neoplasms. *PLoS ONE* **2017**, *12*, e0188876. [CrossRef]
19. Er, L.-M.; Li, Y.; Wu, M.-L.; Zhao, Q.; Tan, B.-B.; Wang, X.-L.; Wang, S.-J. Expression of IMP3 as a marker for predicting poor outcome in gastroenteropancreatic neuroendocrine neoplasms. *Oncol. Lett.* **2017**, *13*, 2391–2396. [CrossRef]
20. Fottner, C.; Ferrata, M.; Weber, M.M. Hormone secreting gastro-entero-pancreatic neuroendocrine neoplasias (GEP-NEN): When to consider, how to diagnose? *Rev. Endocr. Metab. Disord.* **2017**, *18*, 393–410. [CrossRef]
21. Seymour, N.; Sawh, S.C. Mega-dose intravenous octreotide for the treatment of carcinoid crisis: A systematic review. *Can. J. Anesth.* **2013**, *60*, 492–499. [CrossRef] [PubMed]
22. Sundin, A.; Arnold, R.; Baudin, E.; Cwikla, J.; Eriksson, B.; Fanti, S.; Fazio, N.; Giammarile, F.; Hicks, R.; Kjaer, A.; et al. ENETS Consensus Guidelines for the Standards of Care in Neuroendocrine Tumors: Radiological, Nuclear Medicine and Hybrid Imaging. *Neuroendocrinology* **2017**, *105*, 212–244. [CrossRef] [PubMed]
23. Jin, X.-F.; Spampatti, M.P.; Spitzweg, C.; Auernhammer, C.J. Supportive therapy in gastroenteropancreatic neuroendocrine tumors: Often forgotten but important. *Rev. Endocr. Metab. Disord.* **2018**, *19*, 145–158. [CrossRef] [PubMed]

24. Garcia-Carbonero, R.; Sorbye, H.; Baudin, E.; Raymond, E.; Wiedenmann, B.; Niederle, B.; Sedlackova, E.; Toumpanakis, C.; Anlauf, M.; Cwikla, J.; et al. Consensus Guidelines for High Grade Gastro-Entero-Pancreatic (GEP) Neuroendocrine Tumours and Neuroendocrine Carcinomas (NEC). *Neuroendocrinology* **2016**, *103*, 186–194. [CrossRef]
25. Cavalcanti, M.S.; Gönen, M.; Klimstra, D.S. The ENETS/WHO grading system for neuroendocrine neoplasms of the gastroenteropancreatic system: A review of the current state, limitations and proposals for modifications. *Int. J. Endocr. Oncol.* **2016**, *3*, 203–219. [CrossRef]
26. Niederle, B.; Pape, U.-F.; Costa, F.; Gross, D.; Kelestimur, F.; Knigge, U.; Öberg, K.; Pavel, M.; Perren, A.; Toumpanakis, C.; et al. ENETS Consensus Guidelines Update for Neuroendocrine Neoplasms of the Jejunum and Ileum. *Neuroendocrinology* **2016**, *103*, 125–138. [CrossRef]
27. Inzani, F.; Petrone, G.; Rindi, G. The New World Health Organization Classification for Pancreatic Neuroendocrine Neoplasia. *Endocrinol. Metab. Clin. N. Am.* **2018**, *47*, 463–470. [CrossRef]
28. Ito, T.; Igarashi, H.; Uehara, H.; Berna, M.J.; Jensen, R.T. Causes of Death and Prognostic Factors in Multiple Endocrine Neoplasia Type 1. *Medicine* **2013**, *92*, 135–181. [CrossRef]
29. Kos-Kudła, B.; Blicharz-Dorniak, J.; Strzelczyk, J.; Bałdys-Waligórska, A.; Bednarczuk, T.; Bolanowski, M.; Boratyn-Nowicka, A.; Borowska, M.; Cichocki, A.; Cwikla, J.; et al. Diagnostic and therapeutic guidelines for gastro-entero-pancreatic neuroendocrine neoplasms (recommended by the Polish Network of Neuroendocrine Tumours). *Endokrynol Pol.* **2017**, *68*, 79–110. [CrossRef]
30. Modrzejewski, A.; Śledź, M.; Chęciński, P.; Parafiniuk, M.; Pawlik, A.; Kurzawski, M.; Czerny, B. Carcinoid tumor of the gallbladder: Laparoscopic resection and review of the literature. Videosurgery and other miniinvasive techniques. *Surgery* **2009**, *4*, 72–75.
31. Yao, J.C.; Hassan, M.; Phan, A.; Dagohoy, C.; Leary, C.; Mares, J.E.; Abdalla, E.K.; Fleming, J.B.; Vauthey, J.-N.; Rashid, A.; et al. One Hundred Years After "Carcinoid": Epidemiology of and Prognostic Factors for Neuroendocrine Tumors in 35,825 Cases in the United States. *J. Clin. Oncol.* **2008**, *26*, 3063–3072. [CrossRef] [PubMed]
32. Pandit, S.; Bhusal, K. Carcinoid Syndrome. In *StatPearls [Internet]*; 2018. Available online: https://www.ncbi.nlm.nih.gov/books/NBK448096/ (accessed on 1 January 2020).
33. Bhattacharya, S. Risk factors for the development and progression of carcinoid heart disease. *Am. J. Cardiol.* **2011**, *107*, 1221–1226. [CrossRef] [PubMed]
34. Sagar, V.M.; Cooper, S.C.; Johnson, J.; Shetty, S.; Shah, T. Gastrointestinal manifestations of neuroendocrine tumours: Their investigation and management. *Postgrad. Med. J.* **2017**, *93*, 494–497. [CrossRef] [PubMed]
35. Gierach, M.; Gierach, J.; Skowrońska, A.; Junik, R. Neuroendocrine tumors—Insulinoma in clinical practice. *Postępy Nauk Med.* **2013**, *12*, 906–909.
36. Epelboym, I.; Mazeh, H. Zollinger-Ellison Syndrome: Classical Considerations and Current Controversies. *Oncologist* **2013**, *19*, 44–50. [CrossRef]
37. Cingam, S.R.; Karanchi, H. Gastrinoma. Copyright© 2018, StatPearls Publishing LLC. Available online: https://www.ncbi.nlm.nih.gov/books/NBK441842/ (accessed on 1 January 2020).
38. Sandhu, S.; Jialal, I. VIPoma. Copyright© 2018, StatPearls Publishing LLC. Available online: https://www.ncbi.nlm.nih.gov/books/NBK507698/ (accessed on 1 January 2020).
39. John, A.; Schwartz, R.A. Glucagonoma syndrome: A review and update on treatment. *J. Eur. Acad. Dermatol. Venereol.* **2016**, *30*, 2016–2022. [CrossRef]
40. Vinink, A.; Pacak, K.; Feliberti, E.; Perry, R.R. Glucagonoma Syndrome. Copyright 2000–2018, MDText.com, Inc. Available online: https://www.ncbi.nlm.nih.gov/books/NBK279041/ (accessed on 1 January 2020).
41. Vinik, A.; Pacak, K.; Feliberti, E.; Perry, R.R. Somatostatinoma. Copyright 2000-2018, MDText.com, Inc. Available online: https://www.ncbi.nlm.nih.gov/books/NBK279034/ (accessed on 1 January 2020).
42. Garbrecht, N.; Anlauf, M.; Schmitt, A.; Henopp, T.; Sipos, B.; Raffel, A.; Eisenberger, C.F.; Knoefel, W.T.; Pavel, M.; Fottner, C.; et al. Somatostatin-producing neuroendocrine tumors of the duodenum and pancreas: Incidence, types, biological behavior, association with inherited syndromes, and functional activity. *Endocr. Relat. Cancer* **2008**, *15*, 229–241. [CrossRef]
43. Falconi, M.; Eriksson, B.; Kaltsas, G.; Bartsch, D.; Capdevila, J.; Caplin, M.; Kos-Kudła, B.; Kwekkeboom, D.; Rindi, G.; Kloppel, G.; et al. ENETS Consensus Guidelines Update for the Management of Patients with Functional Pancreatic Neuroendocrine Tumors and Non-Functional Pancreatic Neuroendocrine Tumors. *Neuroendocrinology* **2016**, *103*, 153–171. [CrossRef]

44. Coriat, R.; Walter, T.; Terris, B.; Couvelard, A.; Ruszniewski, P. Gastroenteropancreatic Well-Differentiated Grade 3 Neuroendocrine Tumors: Review and Position Statement. *Oncologist* **2016**, *21*, 1191–1199. [CrossRef]
45. Watzka, F.M.; Fottner, C.; Miederer, M.; Weber, M.M.; Schad, A.; Lang, H.; Musholt, T.J. Surgical Treatment of NEN of Small Bowel: A Retrospective Analysis. *World J. Surg.* **2016**, *40*, 749–758. [CrossRef]
46. Jiang, S.-H.; Li, J.; Dong, F.; Yang, J.-Y.; Liu, D.-J.; Yang, X.; Wang, Y.-H.; Yang, M.; Fu, X.; Zhang, X.-X.; et al. Increased Serotonin Signaling Contributes to the Warburg Effect in Pancreatic Tumor Cells Under Metabolic Stress and Promotes Growth of Pancreatic Tumors in Mice. *Gastroenterology* **2017**, *153*, 277–291.e19. [CrossRef] [PubMed]
47. Khan, M.S.; Caplin, E.M. Therapeutic management of patients with gastroenteropancreatic neuroendocrine tumours. *Endocr. Relat. Cancer* **2011**, *18*, S53–S74. [CrossRef] [PubMed]
48. Appetecchia, M.; Baldelli, R. Somatostatin analogues in the treatment of gastroenteropancreatic neuroendocrine tumours, current aspects and new perspectives. *J. Exp. Clin. Cancer Res.* **2010**, *29*, 19. [CrossRef] [PubMed]
49. Rinke, A.; Müller, H.-H.; Schade-Brittinger, C.; Klose, K.-J.; Barth, P.; Wied, M.; Mayer, C.; Aminossadati, B.; Pape, U.-F.; Bläker, M.; et al. Placebo-Controlled, Double-Blind, Prospective, Randomized Study on the Effect of Octreotide LAR in the Control of Tumor Growth in Patients With Metastatic Neuroendocrine Midgut Tumors: A Report From the PROMID Study Group. *J. Clin. Oncol.* **2009**, *27*, 4656–4663. [CrossRef]
50. Caplin, M.E.; Pavel, M.; Cwikla, J.; Phan, A.T.; Raderer, M.; Sedlackova, E.; Cadiot, G.; Wolin, E.M.; Capdevila, J.; Wall, L.; et al. Lanreotide in Metastatic Enteropancreatic Neuroendocrine Tumors. *N. Engl. J. Med.* **2014**, *371*, 224–233. [CrossRef]
51. Faivre, S.; Djelloul, S.; Raymond, E. New Paradigms in Anticancer Therapy: Targeting Multiple Signaling Pathways With Kinase Inhibitors. *Semin. Oncol.* **2006**, *33*, 407–420. [CrossRef]
52. Uri, I.; Avniel-Polak, S.; Gross, D.J.; Grozinsky-Glasberg, S. Update in the Therapy of Advanced Neuroendocrine Tumors. *Curr. Treat. Options Oncol.* **2017**, *18*, 72. [CrossRef]
53. Bednarczuk, T.; Bolanowski, M.; Zemczak, A.; Bałdys-Waligórska, A.; Blicharz-Dorniak, J.; Boratyn-Nowicka, A.; Borowska, M.; Cichocki, A.; Ćwikła, J.B.; Falconi, M.; et al. Neuroendocrine neoplasms of the small intestine and appendicitis - principles of conduct (recommended by the Polish Neuroendocrine Tumor Network). *Endokrynol. Pol.* **2013**, *64*, 480–493.
54. Yao, J.C.; Fazio, N.; Singh, S.; Buzzoni, R.; Carnaghi, C.; Wolin, E.; Tomasek, J.; Raderer, M.; Lahner, H.; Voi, M.; et al. Everolimus for the treatment of advanced, non-functional neuroendocrine tumours of the lung or gastrointestinal tract (RADIANT-4): A randomised, placebo-controlled, phase 3 study. *Lancet* **2015**, *387*, 968–977. [CrossRef]
55. Grillo, F.; Florio, T.; Ferraù, F.; Kara, E.; Fanciulli, G.; Faggiano, A.; Colao, A.A.L. Emerging multitarget tyrosine kinase inhibitors in the treatment of neuroendocrine neoplasms. *Endocr. Relat. Cancer* **2018**, *25*, R453–R466. [CrossRef]
56. Raymond, E.; Dahan, L.; Raoul, J.-L.; Bang, Y.-J.; Borbath, I.; Lombard-Bohas, C.; Valle, J.W.; Metrakos, P.; Smisth, D.; Vinik, A.; et al. Sunitinib Malate for the Treatment of Pancreatic Neuroendocrine Tumors. *N. Engl. J. Med.* **2011**, *364*, 501–513. [CrossRef] [PubMed]
57. Pavel, E.M.; Gross, D.J.; Benavent, M.; Perros, P.; Srirajaskanthan, R.; Warner, R.R.P.; Kulke, M.H.; Anthony, L.B.; Kunz, P.L.; Horsch, D.; et al. Telotristat ethyl in carcinoid syndrome: Safety and efficacy in the TELECAST phase 3 trial. *Endocr.-Relat. Cancer* **2018**, *25*, 309–322. [CrossRef] [PubMed]
58. Yao, J.C.; Pavel, M.; Lombard-Bohas, C.; Van Cutsem, E.; Voi, M.; Brandt, U.; He, W.; Chen, D.; Capdevila, J.; De Vries, E.G.; et al. Everolimus for the Treatment of Advanced Pancreatic Neuroendocrine Tumors: Overall Survival and Circulating Biomarkers From the Randomized, Phase III RADIANT-3 Study. *J. Clin. Oncol.* **2016**, *34*, 3906–3913. [CrossRef]
59. Saif, M.W.; Kaley, K.; Brennan, M.; Garcon, M.C.; Rodriguez, G.; Rodriguez, T. A retrospective study of capecitabine/temozolomide (CAPTEM) regimen in the treatment of metastatic pancreatic neuroendocrine tumors (pNETs) afer failing previous therapy. *J. Pancreas* **2013**, *14*, 498–501.
60. Sahu, A.; Jefford, M.; Lai-Kwon, J.; Thai, A.; Hicks, R.J.; Michael, M. CAPTEM in Metastatic Well-Differentiated Intermediate to High Grade Neuroendocrine Tumors: A Single Centre Experience. *J. Oncol.* **2019**, *2019*, 9032753. [CrossRef]

61. Pavel, M.; O'Toole, D.; Costa, F.; Capdevila, J.; Gross, D.; Kianmanesh, R.; Krenning, E.; Knigge, U.; Salazar, R.; Pape, U.-F.; et al. ENETS Consensus Guidelines Update for the Management of Distant Metastatic Disease of Intestinal, Pancreatic, Bronchial Neuroendocrine Neoplasms (NEN) and NEN of Unknown Primary Site. *Neuroendocrinology* **2016**, *103*, 172–185. [CrossRef]
62. Kaderli, R.M.; Spanjol, M.; Kollár, A.; Bütikofer, L.; Gloy, V.; Dumont, R.A.; Seiler, C.A.; Christ, E.R.; Radojewski, P.; Briel, M.; et al. Therapeutic Options for Neuroendocrine Tumors. *JAMA Oncol.* **2019**, *5*, 480. [CrossRef]
63. Liu, T.; Liao, J.; Dang, J.; Li, G. Treatments for patients with advanced neuroendocrine tumors: A network meta-analysis. *Ther. Adv. Med. Oncol.* **2019**, *11*. [CrossRef]
64. Berardi, R.; Rinaldi, S.; Torniai, M.; Morgese, F.; Partelli, S.; Caramanti, M.; Onofri, A.; Polenta, V.; Pagliaretta, S.; Falconi, M.; et al. Gastrointestinal neuroendocrine tumors: Searching the optimal treatment strategy—A literature review. *Crit. Rev. Oncol.* **2016**, *98*, 264–274. [CrossRef]
65. Shachar, S.S.; Williams, G.R. The Obesity Paradox in Cancer-Moving Beyond BMI. *Cancer Epidemiol. Biomark. Prev.* **2017**, *26*, 13–16. [CrossRef]
66. Sharpless, N.E.; DePinho, R.A. p53: Good cop/bad cop. *Cell* **2002**, *110*, 9–12. [CrossRef]
67. Fontana, L.; Klein, S. Aging, Adiposity, and Calorie Restriction. *JAMA* **2007**, *297*, 986. [CrossRef] [PubMed]
68. Surh, Y.-J. Cancer chemoprevention with dietary phytochemicals. *Nat. Rev. Cancer* **2003**, *3*, 768–780. [CrossRef] [PubMed]
69. Go, V.; Srihari, P.; Burns, L.A.K. Nutrition and Gastroenteropancreatic Neuroendocrine Tumors. *Endocrinol. Metab. Clin. N. Am.* **2010**, *39*, 827–837. [CrossRef]
70. Barrea, L.; Altieri, B.; Muscogiuri, G.; Laudisio, D.; Annunziata, G.; Colao, A.; Faggiano, A.; Savastano, S. Impact of Nutritional Status on Gastroenteropancreatic Neuroendocrine Tumors (GEP-NET) Aggressiveness. *Nutrients* **2018**, *10*, 1854. [CrossRef]
71. Wang, X.; Ouyang, Y.; Liu, J.; Zhu, M.; Zhao, G.; Bao, W.; Hu, F.B. Fruit and vegetable consumption and mortality from all causes, cardiovascular disease, and cancer: Systematic review and dose-response meta-analysis of prospective cohort studies. *BMJ* **2014**, *349*, g4490. [CrossRef]
72. De Pergola, G.; Silvestris, F. Obesity as a Major Risk Factor for Cancer. *J. Obes.* **2013**, *2013*, 1–11. [CrossRef]
73. Herrera-Martínez, A.D.; Hofland, J.; Hofland, L.J.; Brabander, T.; Eskens, F.A.L.M.; Moreno, M.A.G.; Luque, R.M.; Castaño, J.P.; De Herder, W.W.; Feelders, R.A. Targeted Systemic Treatment of Neuroendocrine Tumors: Current Options and Future Perspectives. *Drugs* **2018**, *79*, 21–42. [CrossRef]
74. Altieri, B.; Barrea, L.; Modica, R.; Muscogiuri, G.; Savastano, S.; Colao, A.; Faggiano, A. Nutrition and neuroendocrine tumors: An update of the literature. *Rev. Endocr. Metab. Disord.* **2018**, *19*, 159–167. [CrossRef]
75. Longo, V.D.; Fontana, L. Calorie restriction and cancer prevention: Metabolic and molecular mechanisms. *Trends Pharmacol. Sci.* **2010**, *31*, 89–98. [CrossRef]
76. Turati, F.; Rossi, M.; Pelucchi, C.; Levi, F.; La Vecchia, C. Fruit and vegetables and cancer risk: A review of southern European studies. *Br. J. Nutr.* **2015**, *113*, S102–S110. [CrossRef] [PubMed]
77. Promotion. USDoHaHSOoDPaH. 2015–2020 Dietary Guidelines for Americans. Available online: https://health.gov/dietaryguidelines/?_ga=2.93506781.1509651194.1526503567-2137365160.1526503567 (accessed on 1 January 2020).
78. Blanchard, C.M.; Courneya, K.S.; Stein, K. Cancer Survivors' Adherence to Lifestyle Behavior Recommendations and Associations With Health-Related Quality of Life: Results From the American Cancer Society's SCS-II. *J. Clin. Oncol.* **2008**, *26*, 2198–2204. [CrossRef] [PubMed]
79. Dietary Guidelines Advisory Committee; US Department of Health and Human Services; US Department of Agriculture. *Dietary Guidelines for Americans 2005*; US Government Printing Office: Washington, DC, USA, 2005.
80. Özoğul, F.; Hamed, I. The importance of lactic acid bacteria for the prevention of bacterial growth and their biogenic amines formation: A review. *Crit. Rev. Food Sci. Nutr.* **2017**, *58*, 1660–1670. [CrossRef] [PubMed]
81. Jiang, W.; Xu, Y.; Li, C.; Dong, X.; Wang, N. Biogenic amines in commercially produced Yulu, a Chinese fermented fish sauce. *Food Addit. Contam. Part B* **2013**, *7*, 25–29. [CrossRef]
82. Fan, P.; Song, P.; Li, L.; Huang, C.; Chen, J.; Yang, W.; Qiao, S.; Wu, G.; Zhang, G.; Ma, X. Roles of Biogenic Amines in Intestinal Signaling. *Curr. Protein Pept. Sci.* **2017**, *18*, 532–540. [CrossRef]
83. Tuberoso, C.; Serreli, G.; Montoro, P.; D'Urso, G.; Congiu, F.; Kowalczyk, A. Biogenic amines and other polar compounds in long aged oxidized Vernaccia di Oristano white wines. *Food Res. Int.* **2018**, *111*, 97–103. [CrossRef]

84. Martin, I.S.M.; Brachero, S.; Vilar, E.G. Histamine intolerance and dietary management: A complete review. *Allergol. Immunopathol.* **2016**, *44*, 475–483. [CrossRef]
85. Wüthrich, B. Allergic and intolerance reactions to wine. *Allergol. Sel.* **2018**, *2*, 80–88. [CrossRef]
86. Riederer, P.; Müller, T. Use of monoamine oxidase inhibitors in chronic neurodegeneration. *Expert Opin. Drug Metab. Toxicol.* **2017**, *13*, 233–240. [CrossRef]
87. Schink, M.; Konturek, P.C.; Tietz, E.; Dieterich, W.; Pinzer, T.C.; Wirtz, S.; Neurath, M.F.; Zopf, Y. Microbial Patterns in Patients with Histamine Intolerance. *J. Physiol. Pharmacol.* **2018**, *69*, 579–593.
88. Latorre-Moratalla, M.; Comas-Basté, O.; Bover-Cid, S.; Vidal-Carou, M.C. Tyramine and histamine risk assessment related to consumption of dry fermented sausages by the Spanish population. *Food Chem. Toxicol.* **2017**, *99*, 78–85. [CrossRef] [PubMed]
89. Ercan, S.Ş.; Bozkurt, H.; Soysal, Ç. Significance of Biogenic Amines in Foods and Their Reduction Methods. *J. Food Sci. Eng.* **2013**, *3*, 395–410. [CrossRef]
90. Franke, G.; Cwiková, O. Biogenic amines in smear ripened cheeses. *Potravin. Slovak J. Food Sci.* **2019**, *13*, 378–384. [CrossRef]
91. Naraev, B.G.; Halland, M.; Halperin, D.M.; Purvis, A.J.; O'dorisioT, M.; Halfdanarson, T.R. Management of Diarrhea in Patients With Carcinoid Syndrome. *Pancreas* **2019**, *48*, 961–972. [CrossRef]
92. Clement, D.S.V.M.; Tesselaar, M.E.T.; E Van Leerdam, M.; Srirajaskanthan, R.; Ramage, J. Nutritional and vitamin status in patients with neuroendocrine neoplasms. *World J. Gastroenterol.* **2019**, *25*, 1171–1184. [CrossRef]
93. Laing, E.; Kiss, N.; Michael, M.; Gough, K.; Krishnasamy, M.; Whyand, T.; Auer, B. Investigating Nutrition-Related Complications and Quality of Life in Patients With Gastroenteropancreatic Neuroendocrine Tumors: Protocol for a Mixed-Methods Prospective Study. *JMIR Res. Protoc.* **2018**, *7*, e11228. [CrossRef]
94. Hutton, J.L.; Martin, L.; Field, C.J.; Wismer, W.V.; Bruera, E.D.; Watanabe, S.M.; Baracos, V.E. Dietary patterns in patients with advanced cancer: Implications for anorexia-cachexia therapy. *Am. J. Clin. Nutr.* **2006**, *84*, 1163–1170. [CrossRef]
95. Shaw, C.; Taylor, L. Treatment-Related Diarrhea in Patients with Cancer. *Clin. J. Oncol. Nurs.* **2012**, *16*, 413–417. [CrossRef]
96. Jones, L.W.; Demark-Wahnefried, W. Diet, exercise, and complementary therapies after primary treatment for cancer. *Lancet Oncol.* **2006**, *7*, 1017–1026. [CrossRef]
97. Paris, S.; García, M.; Trufero, M.; Sorrosal, L.; Gracia, C.; Alaminos, L.; Sanz-París, A.; Martinez-García, M.; Trufero, J.M.; Lambea, J.; et al. Oligomeric Enteral Nutrition in Undernutrition, due to Oncology Treatment-Related Diarrhea. Systematic Review and Proposal of An Algorithm of Action. *Nutrients* **2019**, *11*, 1888. [CrossRef]
98. Thompson, K.L.; Elliott, L.; Fuchs-Tarlovsky, V.; Levin, R.; Voss, A.C.; Piemonte, T. Oncology Evidence-Based Nutrition Practice Guideline for Adults. *J. Acad. Nutr. Diet.* **2017**, *117*, 297–310.e47. [CrossRef] [PubMed]
99. Wedlake, L.J.; Shaw, C.; Whelan, K.; Andreyev, H.J.N. Systematic review: The efficacy of nutritional interventions to counteract acute gastrointestinal toxicity during therapeutic pelvic radiotherapy. *Aliment. Pharmacol. Ther.* **2013**, *37*, 1046–1056. [CrossRef] [PubMed]
100. Mardas, M.; Madry, R.; Stelmach-Mardas, M. Link between diet and chemotherapy related gastrointestinal side effects. *Contemp. Oncol.* **2017**, *21*, 162–167. [CrossRef] [PubMed]
101. Grabenbauer, G.G.; Holger, G. Management of radiation and chemotherapy related acute toxicity in gastrointestinal cancer. *Best Pract. Res. Clin. Gastroenterol.* **2016**, *30*, 655–664. [CrossRef] [PubMed]
102. De Hosson, L.D.; Stelwagen, J.; Bouma, G.; Sijtema, B.; Huitema, S.; Van Faassen, H.J.R.; De Bock, G.H.; A De Groot, D.J.; E Campmans-Kuijpers, M.J.; Kema, I.P.; et al. Towards optimal personalized diet and vitamin supplementation in NET patients. *Endocr. Relat. Cancer* **2018**, *25*, L23–L26. [CrossRef] [PubMed]
103. Yoon, S.R.; Lee, J.H.; Lee, J.H.; Na, G.Y.; Lee, K.-H.; Lee, Y.-B.; Jung, G.-H.; Kim, O.Y. Low-FODMAP formula improves diarrhea and nutritional status in hospitalized patients receiving enteral nutrition: A randomized, multicenter, double-blind clinical trial. *Nutr. J.* **2015**, *14*, 116. [CrossRef] [PubMed]
104. Gibson, R.; Keefe, D.M.K. Cancer chemotherapy-induced diarrhoea and constipation: Mechanisms of damage and prevention strategies. *Support. Care Cancer* **2006**, *14*, 890–900. [CrossRef]
105. Vandeputte, D.; Falony, G.; Vieira-Silva, S.; Wang, J.; Sailer, M.; Theis, S.; Verbeke, K.; Raes, J. Prebiotic inulin-type fructans induce specific changes in the human gut microbiota. *Gut* **2017**, *66*, 1968–1974. [CrossRef]

106. Riezzo, G.; Orlando, A.; D'Attoma, B.; Linsalata, M.; Martulli, M.; Russo, F. Randomised double blind placebo controlled trial on Lactobacillus reuteri DSM 17938: Improvement in symptoms and bowel habit in functional constipation. *Benef. Microbes* **2018**, *9*, 51–60. [CrossRef]
107. Wołkowicz, T.; Januszkiewicz, A.; Szych, J. Microbiome of the digestive tract and its dysbiosis as an important factor affecting the health condition of the human body. *Med. Dośw. Mikrobiol.* **2014**, *66*, 223–235.
108. Bouma, G.; van Faassen, M.; Kats-Ugurlu, G.; Vries de, E.G.E.; Kema, I.P.; Walenkamp, A.M.E. Niacin (Vitamin B-3) supplementation in patients with serotonin-producing neuroendocrine tumor. *Neuroendocrinology* **2016**, *103*, 489–494. [CrossRef] [PubMed]
109. Shah, G.M.; Shah, R.G.; Veillette, H.; Kirkland, J.B.; Pasieka, J.L.; Warner, R.R.P. Biochemical Assessment of Niacin Deficiency Among Carcinoid Cancer Patients. *Am. J. Gastroenterol.* **2005**, *100*, 2307–2314. [CrossRef] [PubMed]
110. Brennan, G.T.; Saif, M.W. Pancreatic Enzyme Replacement Therapy: A Concise Review. *J. Pancreas* **2019**, *20*, 126–129.
111. Qureshi, S.A.; Burch, N.; Druce, M.; Hattersley, A.; Khan, S.; Gopalakrishnan, K.; Darby, C.; Wong, J.L.H.; Davies, L.; Fletcher, S.; et al. Screening for malnutrition in patients with gastro-entero-pancreatic neuroendocrine tumours: A cross-sectional study. *BMJ Open* **2016**, *6*, e010765. [CrossRef] [PubMed]
112. Arends, J.; Bodoky, G.; Bozzetti, F.; Fearon, K.; Muscaritoli, M.; Selga, G.; Schueren, M.V.B.-D.V.D.; Von Meyenfeldt, M.; Zürcher, G.; Fietkau, R.; et al. ESPEN Guidelines on Enteral Nutrition: Non-surgical oncology. *Clin. Nutr.* **2006**, *25*, 245–259. [CrossRef]
113. Elting, L.S.; Cooksley, C.; Chambers, M.; Cantor, S.B.; Manzullo, E.; Rubenstein, E.B. The burdens of cancer therapy. *Cancer* **2003**, *98*, 1531–1539. [CrossRef]
114. Borre, M.; Dam, G.A.; Knudsen, A.W.; Grønbæk, H. Nutritional status and nutritional risk in patients with neuroendocrine tumors. *Scand. J. Gastroenterol.* **2018**, *53*, 284–292. [CrossRef]
115. Robbins, H.L.; Symington, M.; Mosterman, B.; Goodby, J.; Davies, L.; Dimitriadis, G.K.; Kaltsas, G.A.; Randeva, H.S.; Weickert, M.O. Supplementation of Vitamin D Deficiency in Patients with Neuroendocrine Tumors Using Over-the-Counter Vitamin D3 Preparations. *Nutr. Cancer* **2018**, *70*, 748–754. [CrossRef]
116. Lavriv, D.S.; Neves, P.; Ravasco, P. Should omega-3 fatty acids be used for adjuvant treatment of cancer cachexia? *Clin. Nutr. ESPEN* **2018**, *25*, 18–25. [CrossRef]
117. Keshet, R.; Erez, A. Arginine and the metabolic regulation of nitric oxide synthesis in cancer. *Dis. Model. Mech.* **2018**, *11*, dmm033332. [CrossRef]
118. Maasberg, S.; Knappe-Drzikova, B.; Vonderbeck, D.; Jann, H.; Weylandt, K.; Grieser, C.; Pascher, A.; Schefold, J.C.; Pavel, M.; Wiedenmann, B.; et al. Malnutrition Predicts Clinical Outcome in Patients with Neuroendocrine Neoplasia. *Neuroendocrinology* **2015**, *104*, 11–25. [CrossRef] [PubMed]
119. Lamarca, A.; McCallum, L.; Nuttall, C.; Barriuso, J.; Backen, A.; Frizziero, M.; Leon, R.; Mansoor, W.; McNamara, M.G.; Hubner, R.A.; et al. Somatostatin analogue-induced pancreatic exocrine insufficiency in patients with neuroendocrine tumors: Results of a prospective observational study. *Expert Rev. Gastroenterol. Hepatol.* **2018**, *12*, 723–731. [CrossRef] [PubMed]

nutrients

MDPI

Article

Malnutrition Screening and Assessment in the Cancer Care Ambulatory Setting: Mortality Predictability and Validity of the Patient-Generated Subjective Global Assessment Short form (PG-SGA SF) and the GLIM Criteria

Lynette M. De Groot [1,2], Gahee Lee [3], Antoinette Ackerie [1] and Barbara S. van der Meij [1,2,3,*]

1 Dietetics and Foodservices, Mater Health, Brisbane, 4101 QLD, Australia; lynette.degroot@gmail.com (L.M.D.G.); antoinette.ackerie@mater.org.au (A.A.)

2 Mater Research Institute, University of Queensland, Brisbane, 4101 QLD, Australia

3 Bond University Nutrition and Dietetics Research Group, Faculty of Health Sciences and Medicine, Bond University, Gold Coast, 4226 QLD, Australia; gahee.lee@student.bond.edu.au

* Correspondence: bvanderm@bond.edu.au

Received: 17 June 2020; Accepted: 21 July 2020; Published: 30 July 2020

Abstract: Background: A valid malnutrition screening tool (MST) is essential to provide timely nutrition support in ambulatory cancer care settings. The aim of this study is to investigate the validity of the Patient-Generated Subjective Global Assessment Short Form (PG-SGA SF) and the new Global Leadership Initiative on Malnutrition (GLIM) criteria as compared to the reference standard, the Patient-Generated Subjective Global Assessment (PG-SGA). Methods: Cross-sectional observational study including 246 adult ambulatory patients with cancer receiving in-chair intravenous treatment at a cancer care centre in Australia. Anthropometrics, handgrip strength and patient descriptive data were assessed. Nutritional risk was identified using MST and PG-SGA SF, nutritional status using PG-SGA and GLIM. Sensitivity (Se), specificity (Sp), positive and negative predictive values and kappa (k) were analysed. Associations between malnutrition and 1-year mortality were investigated by Cox survival analyses. Results: A PG-SGA SF cut-off score ≥5 had the highest agreement when compared with the PG-SGA (Se: 89%, Sp: 80%, k = 0.49, moderate agreement). Malnutrition risk (PG-SGA SF ≥ 5) was 31% vs. 24% (MST). For malnutrition according to GLIM, the Se was 76% and Sp was 73% (k = 0.32, fair agreement) when compared to PG-SGA. The addition of handgrip strength to PG-SGA SF or GLIM did not improve Se, Sp or agreement. Of 100 patients who provided feedback, 97% of patients found the PG-SGA SF questions easy to understand, and 81% reported that it did not take too long to complete. PG-SGA SF ≥ 5 and severe malnutrition by GLIM were associated with 1-year mortality risk. Conclusions: The PG-SGA SF and GLIM criteria are accurate, sensitive and specific malnutrition screening and assessment tools in the ambulatory cancer care setting. The addition of handgrip strength tests did not improve the recognition of malnutrition or mortality risk.

Keywords: malnutrition; cancer; handgrip strength; nutrition assessment

---

1. Introduction

In 2019, it is estimated that almost 145,000 new cases of cancer will be diagnosed in Australia [1]. Throughout their cancer journey, around 30–90% of patients experience malnutrition [2,3]. This is concerning, as malnutrition is associated with reduced treatment effectiveness [4–6], functional status [4,6], quality of life [4,7] and survival [4–6,8]. Research suggests that early nutrition intervention may prevent nutritional deterioration in patients with cancer [9–11] and as such it is therefore recommended that regular nutrition screening to facilitate timely nutrition care occurs for all patients undergoing cancer treatment [12].

Patients with cancer often receive treatment in outpatient or ambulatory day care settings. To identify malnutrition risk in these patients, it is essential to choose an efficient nutritional screening tool suitable for ambulatory cancer patients. We know that many screening tools can lack clinical information, sensitivity, specificity and/or are not usable and applicable for busy cancer care centres in which there are high turnover rates of outpatients and limited human resources [13].

A number of malnutrition screening tools exist, with the Malnutrition Screening Tool (MST) being the most commonly used in Australia. The MST has a good validity, sensitivity and specificity to identify malnutrition assessed using the Patient-Generated Subjective Global Assessment (PG-SGA) in ambulatory cancer care patients. The PG-SGA is widely used in clinical practice for assessing nutritional status of cancer patients. The Patient-Generated Subjective Global Assessment Short Form (PG-SGA SF), a component of the full PG-SGA, has recently been receiving new attention as a valid screening tool for nutritional risk [14] and meets the professional standard when compared to the PG-SGA [15,16]. The PG-SGA SF retains the patient completed section of weight history, food intake, nutrition symptoms and physical function, and foregoes the remainder. The optimal cut-off value or score to determine malnutrition risk according to the PG-SGA SF differs in the research and depends on the purpose or objectives of the team using the instrument in clinical practice [17]. Since it is designed to be completed by the patient, it is relatively simple to complete, and it reduces time and additional burden to healthcare professionals [15]. Patient-led screening may improve patient autonomy and aligns with the international focus on patient-centred care. Recently, patient-led screening with the MST was found to be reliable and well accepted by patients attending an ambulatory cancer care setting [18].

Interestingly, to standardize the clinical practice of malnutrition diagnosis in clinical settings, the Global Leadership Initiative on Malnutrition (GLIM) recently proposed diagnostic criteria for malnutrition. The GLIM criteria are objective, global and based on consensus. The GLIM includes a combination of phenotypic (percentage weight loss, low body mass index, reduced muscle mass) and etiologic (reduced food intake or assimilation, acute or chronic inflammation) criteria for the diagnosis of malnutrition. The GLIM criteria have not yet been validated for identifying malnutrition in an ambulatory oncology population, nor its predictive value regarding survival in these patients. There is well established evidence of the PG-SGA's ability to predict clinical outcomes including survival with well-nourished oncology patients having longer survival duration than malnourished patients [14,15].

It has also been shown that handgrip strength is reduced in cancer patients with malnutrition [19]. Handgrip strength has been shown to be a prognostic marker and is positively associated with survival duration, for instance, in older patients with cancer [20,21]. As the aim of nutritional therapy is to restore muscle mass and muscle strength, handgrip strength can serve as an additional parameter to improve the recognition of malnutrition risk or malnutrition. Assessment of muscle function using grip strength is recommended as a supportive measure in the GLIM consensus [22].

Therefore, the aim of this study is to (1) evaluate the agreement, sensitivity and specificity of the PG-SGA SF and GLIM criteria to reference standard of PG-SGA, (2) investigate the added value of handgrip strength test to PG-SGA SF and GLIM to recognise malnutrition risk or malnutrition, (3) evaluate the patient experience and ease of completion of the patient-completed PG-SGA SF and (4) investigate the ability of these tools to predict patient outcomes such as mortality and hospitalisation and whether any tool is superior in this ability.

2. Materials and Methods

We conducted a cross-sectional, observational study at the Mater Cancer Care Centre (MCCC) in Brisbane, QLD, Australia over one week in May 2018 and repeated over one week in March 2019. Patient outcome data were collected 12 months post initial data collection. The protocol received was approved as low risk research by the Human Research Ethics Committee of Mater Misericordiae Limited (HREC/18/MHS/101).

2.1. Participants

Patients with cancer aged 18 years or older were included if they were receiving in-chair intravenous treatment at MCCC. People who were receiving treatment for a benign condition (e.g., rheumatoid arthritis), receiving a blood transfusion only, waiting on hospital admission, or who declined to take part in this study were excluded from participation.

2.2. PG-SGA SF

Patients who consented to the study were asked to complete a paper-based PG-SGA SF and received basic verbal instructions (such as 'tick the box that is applicable') by trained staff. The PG-SGA SF consists of four boxes: box 1, questions regarding body weight (scored 0–5); box 2, food intake (score 0–4); box 3, symptoms affecting oral food intake (scored 0–23); and box 4, regarding activities and function. Based on findings from a previous Australian study [16], box 4 questions were excluded from the PG-SGA SF used in this study. Upon completion, the scores of boxes 1, 2 and 3 were totalled by a dietitian.

A short questionnaire to assess ease of completion and time taken to complete the PG-SGA SF was provided to each patient in 2019. The questionnaire was based on previously developed tools [18] and included four questions regarding clarity and understanding of the tool, and time taken to complete the tool.

2.3. MST and PG-SGA

In line with established guidelines [12], all participants were screened for malnutrition using the MST. The MST includes two questions about recent unintentional weight loss and reduced appetite affecting dietary intake [23] with answers generating a numerical score between 0 and 5. Patients with an MST score ≥2 were classified as 'at risk of malnutrition' and were assessed by dietitians using the PG-SGA to determine their degree of malnutrition. The PG-SGA classifies patients into three categories: (A) well-nourished; (B) moderately malnourished; or (C) severely malnourished. For data analysis purposes, patients with MST <2 were assumed as well nourished (PG-SGA A).

2.4. GLIM Criteria

To diagnose malnutrition using the GLIM criteria, weight changes within six months (%) and body mass index (BMI) were calculated using patients' weight history and height (Table 1). Based on the GLIM criteria, participants who had a combination of at least one phenotypic criterion (weight loss and/or low BMI for age) and one etiologic criterion (disease burden and inflammatory condition of cancer) were categorised as malnourished. The remaining participants were categorised as well-nourished.

Table 1. Global Leadership Initiative on Malnutrition (GLIM) phenotypic and etiologic criteria for the diagnosis of malnutrition.

Phenotypic Criteria			Etiologic Criteria	
Weight Loss (%)	**Low Body Mass Index (kg/m²)**	**Reduced Muscle Mass [a]**	**Reduced Food Intake or Assimilation [b,c]**	**Inflammation [d–f]**
>5% within past 6 months or >10% beyond 6 months	<20 if <70 years, or <22 if >70 years Asia: <18.5 if <70 years, or <20 if >70 years	Reduced by validated body composition measuring techniques	≤50% of ER >1 week, or any reduction for >2 weeks, or any chronic GI condition that adversely impacts food assimilation or absorption	Acute disease/injury [d,f] or chronic disease-related [e,f]

ER = energy requirements, GI = gastrointestinal. Requires at least one phenotypic criterion and one etiologic criterion for diagnosis of malnutrition; [a] for example, fat free mass index, by dual-energy absorptiometry or corresponding standards using other body composition methods like bioelectrical impedance analysis, computed tomography (CT) or magnetic resonance imaging (MRI). When not available or by regional preference, physical examination or standard anthropometric measures like mid-arm muscle or calf circumference may be used. Thresholds for reduced muscle mass need to be adapted to race (Asia). Functional assessments like hand-grip strength may be considered as a supportive measure. [b] Consider gastrointestinal symptoms as supportive indicators that can impair food intake or absorption (e.g., dysphagia, nausea, vomiting, diarrhoea, constipation or abdominal pain). Use clinical judgement to discern severity based upon the degree to which intake or absorption are impaired. Symptom intensity, frequency and duration should be noted. [c] Reduced assimilation of food/nutrients is associated with malabsorptive disorders like short bowel syndrome, pancreatic insufficiency and after bariatric surgery. It is also associated with disorders like esophageal stricture, gastroparesis and intestinal pseudo-obstruction. Malabsorption is a clinical diagnosis that manifests as chronic diarrhoea or steatorrhoea. Malabsorption in those with ostomies is evidenced by elevated volumes of output. Use clinical judgement or additional evaluation to discern severity based upon frequency, duration and quantitation of faecal fat and/or volume of losses. [d] Acute disease/injury-related. Severe inflammation is likely to be associated with major infection, burns, trauma or closed head injury. Other acute disease/injury-related conditions are likely to be associated with mild-moderate inflammation. [e] Chronic disease-related conditions. Severe inflammation is not generally associated with chronic disease conditions. Chronic or recurrent mild-moderate inflammation is likely to be associated with malignant disease, chronic obstructive pulmonary disease, congestive heart failure, chronic renal disease or any disease with chronic or recurrent inflammation. Note that transient inflammation of a mild degree does not meet the threshold for this etiologic criterion. [f] C-reactive protein may be used as a supportive laboratory measure.

2.5. Handgrip Strength (HGS)

Handgrip strength tests were performed in the dominant hand using a hydraulic hand dynamometer (Jamar Plus+, Performance Health Supply Inc., Sutton-in-Ashfield, UK). The patient performed the test in a seated position, with the shoulder abducted and neutrally rotated, elbow flexed at 90 degrees, forearm and wrist in a neutral position. Patients were instructed to perform three maximal isometric contractions. Patients took brief pauses between measurements. The maximal value was recorded to the nearest 0.1 kg, and was used to compare with the 10th percentile of age- and gender-dependent reference values [24]. If patients were unable to perform handgrip strength with their dominant hand, the non-dominant hand was used.

2.6. Patient Outcomes

Patient outcome measures were collected 12 months post initial study date and were obtained from the patient electronic medical record. This included mortality, number of hospital admissions during the 12-month period and length of stay of hospital admissions.

2.7. Statistical Analysis

Data analysis was completed using IBM SPSS software V.25.0 for Windows. A significance level of 5% was applied to detect statistical significance. For normally distributed continuous variables, independent t-tests were applied to compare means between malnutrition categories. PG-SGA SF values from 1 to 7 were each analysed to determine optimal cut-off values to determine malnutrition risk for this population. In order to compare the PG-SGA SF and GLIM with the reference instrument (PG-SGA), sensitivity, specificity and positive and negative predictive values for the PG-SGA SF and GLIM against PG-SGA were calculated and receiver operating characteristic (ROC) curves were generated. Kappa coefficient was assessed to investigate the rate of agreement between PG-SGA SF or GLIM and PG-SGA. To determine the validity of the PG-SGA SF, the professional standard 80% for sensitivity and 60% for specificity were determined based on the literature [25,26]. The kappa coefficient was interpreted based on the literature [27,28] as follows: <0 as poor agreement; 0.01–0.20 as slight agreement; 0.21–0.40 as fair agreement; 0.41–0.60 as moderate agreement; 0.61–0.80 as substantial agreement; and 0.81–1.00 as almost perfect agreement. The professional standard for kappa was set to >0.60 [29]. Cox proportional hazards analyses were performed to determine the association between malnutrition diagnoses and mortality 1 year after the audit date. Univariate as well as multivariate analyses adjusted for gender, age (≤65 years vs. >65 years), obesity (BMI ≥ 30 kg/m^2) and diagnosis (breast cancer vs. other types of cancer) were performed on the basis of available literature. The association between malnutrition diagnoses and hospital admissions (Y/N) and length of stay was investigated by logistic and linear regression analysis, respectively (both univariate and multivariate, adjusted for similar confounders).

3. Results

Out of 275 patients that were eligible, a total of 246 patients consented to participate in the study. The mean age was 61.9 ± 13.1 years, and 182 (74%) patients were female. The most common cancer diagnoses were breast (45%), gynaecological (13%) and colorectal (11%) (Table 2).

Table 2. Patient characteristics of 246 patients with cancer attending an ambulatory cancer care centre.

	n (%)
Age [1]	61.9 ± 13.1
Gender	
Male	64 (26)
Female	182 (74)
BMI (WHO categories) [2]	
Underweight (≤18.5 kg/m^2 for <65 years, <24 for ≥65 years)	30 (14)
Healthy weight (18.5–25 kg/m^2 for <65 years, 24–31 for ≥65 years)	102 (47)
Overweight (≥25–30 kg/m^2 for <65 years, >31 for ≥65 years)	53 (24)
Obese (≥30 kg/m^2 for <65 years)	33 (15)
Type of cancer	
Respiratory	21 (9)
Urogenital	15 (6)
Head and neck	1 (1)
Gynaecology	32 (13)
Breast	110 (45)
Haematology	21 (9)
Colorectal	27 (11)
Upper GI	11 (5)
Melanoma	7 (3)
Unknown primary	1 (1)
Emetogenicity of current cancer treatment [3]	
Low	125 (53)
Moderate	64 (27)
High	47 (20)

[1] Average ± SD, [2] n = 218, [3] n = 236. BMI = body mass index; WHO = World Health Organization.

3.1. Validity of Malnutrition Tools/Instruments

The MST had a sensitivity of 100% and a specificity of 90% (kappa: 0.737) when compared to the reference tool PG-SGA.

The sensitivity and specificity for the different cut-off scores and percentiles of PG-SGA SF are depicted in Table 3. The sensitivity was the highest when applying a cut-off score ≥3 for PG-SGA SF, however cut-off scores ≥3, ≥4 and ≥5 all fulfilled the criteria deemed acceptable for validity. A PG-SGA SF cut-off score ≥5 had the highest agreement and deemed most suitable when compared with the reference PG-SGA, with a sensitivity of 89%, a specificity of 80% and a 'moderate agreement' (k = 0.493).

Table 3. Sensitivity and specificity of nutritional indices of malnutrition risk and malnutrition as determined by reference standard Patient-Generated Subjective Global Assessment.

	Sensitivity (%)	Specificity (%)	PPV (%)	NPV (%)	Kappa
Malnutrition Risk					
PG-SGA SF (≥3)	94	62	31	98	0.311
PG-SGA SF (≥4)	92	71	37	98	0.387
PG-SGA SF (≥5)	89	80	45	98	0.493
PG-SGA SF (≥3) and HGS < 10th percentile	21	95	44	87	0.208
PG-SGA SF (≥4) and HGS < 10th percentile	21	96	50	87	0.229
PG-SGA SF (≥5) and HGS < 10th percentile	21	96	50	87	0.229
Malnutrition					
GLIM criteria (moderate, severe or both)	76	73	34	94	0.323
GLIM criteria and HGS < 10th percentile	19	96	43	87	0.186

PG-SGA SF Patient-Generated Subjective Global Assessment Short Form; HGS handgrip strength; GLIM Global Leadership Initiative on Malnutrition; PPV positive predictive value; NPV negative predictive value.

When compared to the reference standard PG-SGA, malnutrition according to the GLIM criteria had a sensitivity of 76%, specificity of 73% and a 'fair agreement' (k = 0.323). When adding handgrip strength < 10th percentile of reference values, sensitivity and specificity for both PG-SGA SF and GLIM declined to around 20% and 95–96%, respectively and kappa declined to approximately 0.2 (poor to slight agreement).

3.2. *Malnutrition Risk and Malnutrition*

According to the MST, 60 patients (24%) were identified as at risk of malnutrition compared to 71 (31%) patients with PG-SGA SF ≥ 5 (Table 4). There were 32 (14%) patients with a handgrip strength (HGS) cut-off score of <10th percentile. When handgrip strength < 10th percentile of reference values was added to PG-SGA SF ≥ 5 the number of patients identified as at risk of malnutrition decreased to 14 (6%).

Table 4. Malnutrition risk as determined by PG-SGA SF.

	At Risk Malnutrition *n* (%)
PG-SGA SF (≥1)	181 (79.4)
PG-SGA SF (≥2)	131 (57)
PG-SGA SF (≥3)	108 (47)
PG-SGA SF (≥4)	90 (39)
PG-SGA SF (≥5)	71 (31)
PG-SGA SF (≥6)	52 (23)
PG-SGA SF (≥7)	41 (18)

PG-SGA SF Patient-Generated Subjective Global Assessment Short Form.

According to the PG-SGA assessed by the dietitian, the number of patients who identified as malnourished was 39 (16%), with 33 (13%) of those moderately (PG-SGA B) malnourished. According to the GLIM criteria for the diagnosis of malnutrition; 77 (35%) out of 220 patients were identified as malnourished (moderately, severely, or both). When handgrip strength was added to the GLIM criteria (HGS < 10th percentile), malnutrition prevalence was reduced to 7% (n = 14).

Malnutrition according to the PG-SGA was identified across all BMI categories, with the lowest percentage of malnutrition in obese patients. Cancer types with the highest rates of malnutrition were respiratory ($n = 21$, 29%), haematology ($n = 21$, 24%) and colorectal cancer ($n = 27$, 22%). In patients receiving chemotherapy agents with a moderate or high emetogenicity risk, more patients were malnourished (24%) compared to patients receiving low emetogenicity risk agents (10%) ($p = 0.002$). The most frequently reported nutrition impact symptom in the malnourished group was 'no appetite' ($n = 23$, 59%), followed by 'things taste funny or have no taste' ($n = 18$, 46%) and 'fatigue' ($n = 12$, 31%).

A reduced food intake in the past month was reported by 66 patients (29.5%). In malnourished patients, a food intake less than usual was reported by 80% ($n = 28$), compared to only 20% ($n = 38$) in well-nourished patients ($p < 0.001$).

3.3. Patient Experience

Table 5 displays data on patients' experiences with self-completing the PG-SGA SF during the 2019 audit. A 95% completion rate was achieved (100 out of 105 questionnaires completed). According to the questionnaire responses, 98% of patients indicated that the instructions to complete the tool were clear ($n = 101$), and most found the questions easy to understand ($n = 101$, 97%). Furthermore, 81% of patients thought the questions did not take too long to complete ($n = 100$) with most reporting it took five minutes or less to complete the PG-SGA SF ($n = 100$, 97%).

Table 5. Patient experience completing PG-SGA SF.

	n	*n* (%)
Questionnaires		
Completed	105	100 (95%)
Not completed		4 (4%)
Partially completed		1 (1%)
The instructions to complete the self-completed tool were clear		
Strongly agree	101	47 (47%)
Agree		52 (52%)
Neutral		1 (1%)
Disagree		1 (1%)
Strongly disagree		0
I understood the questions		
Strongly agree	101	52 (52%)
Agree		46 (46%)
Neutral		2 (2%)
Disagree		1 (1%)
Strongly disagree		0
Please estimate the time taken to complete the questions		
Less than 3 min	100	61 (61%)
3–5 min		36 (36%)
More than 5 min		3 (3%)
The questions took too much time to complete		
Strongly agree	100	6 (6%)
Agree		6 (6%)
Neutral		7 (7%)
Disagree		47 (47%)
Strongly disagree		34 (34%)

3.4. Patient Outcomes

Multivariate analyses showed that malnutrition according to the GLIM criteria was associated with a 2-fold increased 1-year mortality risk, and a 4-fold increased risk when adding handgrip strength lower than the 10th percentile of reference value as a muscle strength parameter. However, these associations were borderline statistically significant (Table 6). Severe malnutrition according to the

GLIM criteria was associated with increased mortality risk (HR 2.9, $p = 0.019$); this was not the case for moderate malnutrition according to GLIM, or for moderate or severe malnutrition (GLIM) when adding handgrip strength < 10th percentile of reference values.

Table 6. Associations between malnutrition diagnoses and 1-year mortality in patients with cancer attending an ambulatory cancer care centre.

Malnutrition Diagnosis	Model 1 [1] HR (95% CI)	Model 2 [1] HR (95% CI)	Model 3 [1] HR (95% CI)
MST ≥ 2	2.554 (1.296; 5.032)	3.021 (1.339; 6.817)	3.392 (1.463; 7.865) $p = 0.004$
PG-SGA B or C	7.128 (2.982; 17.034)	7.305 (2.822; 18.908)	10.373 (3.752; 28.681) $p < 0.001$
GLIM-malnutrition	1.951 (0.930; 4.092)	2.186 (0.982; 4.867)	2.238 (1.004; 4.991) $p = 0.049$
GLIM-malnutrition + handgrip strength < 10th percentile	2.952 (0.856; 10.178)	3.237 (0.882; 11.884)	4.136 (1.009; 16.958) $p = 0.049$
GLIM: moderate malnutrition	1.053 (0.481; 2.304)	1.165 (0.518; 2.620)	1.166 (0.522; 2.606) $p = 0.708$
GLIM: moderate malnutrition + handgrip strength < 10th percentile	4.048 (0.498; 32.928)	3.201 (0.377; 27.205)	3.642 (0.410; 32.349) $p = 0.246$
GLIM: severe malnutrition	2.631 (1.187; 5.831)	2.575 (1.100; 6.031)	2.890 (1.193; 7.001) $p = 0.019$
GLIM: severe malnutrition + handgrip strength < 10th percentile	2.908 (0.849; 9.963)	3.102 (0.854; 11.270)	3.796 (0.948; 15.200) $p = 0.059$
PG-SA SF ≥ 3	3.091 (1.488; 6.422)	3.184 (1.436; 7.060)	3.099 (1.372; 7.000) $p = 0.006$
PG-SA SF ≥ 3 + handgrip strength < 10th percentile	1.477 (0.601; 3.632)	1.547 (0.569; 4.210)	1.570 (0.572; 4.310) $p = 0.381$
PG-SA SF ≥ 5	3.195 (1.560; 6.543)	3.436 (1.535; 7.692)	3.512 (1.483; 8.318) $p = 0.004$
PG-SA SF ≥ 5 + handgrip strength < 10th percentile	1.477 (0.601; 3.632)	1.547 (0.569; 4.210)	1.570 (0.572; 4.310) $p = 0.381$
Handgrip strength < 10th percentile	1.713 (0.735; 3.995)	1.913 (0.721; 5.076)	1.958 (0.729; 5.259) $p = 0.183$

HR hazard ratio; CI confidence interval. [1] Dependent variable: 1-y mortality, independent variables: Model 1: malnutrition diagnosis (0/1), Model 2: malnutrition diagnosis (0/1), gender (female vs. male), age (≥65 vs. <65 y), BMI (≥30 kg/m^2 vs. <30 kg/m^2), Model 3: malnutrition diagnosis (0/1), gender (female vs. male), age (>65 vs. ≤65 y), BMI (≥30 kg/m^2 vs. <30 kg/m^2), breast cancer vs. other types of cancer.

Malnutrition risk according to the PG-SGA SF (both with cut-off ≥3 and ≥5) was associated with a 1.6 or 3.5 increased mortality risk, respectively ($p = 0.006$ and $p = 0.004$). When including handgrip strength, the PG-SGA SF was not associated with 1-year mortality risk.

The reference tools were both associated with 1-year mortality risk: hazard ratio (HR) 3.4 ($p = 0.004$) for MST ≥ 2, and HR 10.4 ($p < 0.001$) for PG-SGA B and PG-SGA C. No associations were observed between number of hospital admissions or total length of stay in hospital during 1-year post initial assessment.

4. Discussion

Findings from this study demonstrate that the PG-SGA SF is an accurate, sensitive and specific malnutrition screening tool in the ambulatory cancer care setting. This is consistent with other studies [14,16] when compared to the full PG-SGA.

Our study identified that cut-off scores of ≥3, ≥4 and ≥5 all fulfilled criteria deemed acceptable for validity, however a cut-off score of ≥5 had the highest agreement with the reference standard of malnutrition according to PG-SGA. Optimal cut-off scores differ according to the literature, reporting ≥3 [16] and ≥6 [15] as optimal cut-off scores. This disparity may be due to different administration methods (patient completed vs. clinician completed). In the current study the patient self-completed the PG-SGA SF. This is more comparable to Gabrielson et al., reporting ≥6 as the optimal cut-off [15]. One reason for our lower cut-off score of ≥5 would be our formatting of the PG-SGA SF, which was the removal of the activity box (box 4). This is a study limitation, however, which accounts for the lower cut-off score.

When compared to MST, the PG-SGA SF identified more patients at risk of malnutrition (31% vs. 24%). This could be explained by the inclusion of nutrition impact symptoms in the PG-SGA SF. Nutrition impact symptoms were common in our cohort, with 29% of patients reporting at least one symptom in the last two weeks. The PG-SGA SF provides valuable information on nutrition impact symptoms, which alone succeeded in identifying patients at risk of malnutrition in previous studies [16,30].

It has been hypothesized that identifying nutrition impact symptoms, especially early in the cancer continuum, may facilitate pro-active malnutrition prevention [14]. For example, patients with no significant weight loss reporting several nutrition impact symptoms are at risk of deterioration in nutritional status and quality of life if they do not receive timely intervention.

Previous studies have indicated that patient-led screening is quick and easy for patients to complete [18]. In our study, patients were provided simple instructions to self-complete the tool, with the addition of patient experience questions in year two of the study. This was included to evaluate the usability and acceptability of the patient completed tool within our local setting. Patient completed PG-SGA SF was well accepted in our study with most indicating the questions easy to understand and that it took five minutes or less to complete the tool. This is similar to a previous study in an ambulatory cancer care setting where they found patient-led screening with the MST was reliable and well accepted by patients [18]. The benefit of the PG-SGA SF over the MST is the additional contributing factors associated with poor dietary intakes, enabling the delivery of the most efficient nutrition intervention [6,31,32].

Malnutrition prevalence in our population differed greatly according to which tool was used, with the GLIM criteria identifying two times the number of patients as malnourished compared to the PG-SGA (35% vs. 16%). PG-SGA is a subjective diagnostic tool that is validated, reliable and widely used in the oncology setting [33–35]. Whilst there is no gold standard for determining nutritional status, for the purpose of our study we used PG-SGA as a primary reference tool to determine the validity of GLIM criteria for malnutrition. When compared to this reference standard PG-SGA, malnutrition according to the GLIM criteria had a sensitivity of 76%, specificity of 73% and a 'fair agreement' ($k = 0.323$). This falls slightly short of the acceptable professional standard of 80% and 60%. The difference in prevalence rates between PG-SGA and GLIM criteria can be theorised by the time frame of involuntary weight loss and the use of BMI. The PG-SGA classification is based on 1-month weight loss percentage, whereas the GLIM criteria include weight loss up to 6 months and beyond 6 months. This may account for the high prevalence according to GLIM, where patients with a history of weight loss that have stabilised weight in the recent period may be classified as not malnourished according to PG-SGA. Additionally, a low BMI as a phenotypic criterion can classify patients as malnourished according to GLIM, whereas BMI is not used in the PG-SGA. There is substantial variation in the use of low BMI as a phenotypic criterion for diagnosis of malnutrition and further research on this is required [36].

Furthermore, in our study a PG-SGA was only completed on patients who scored MST ≥ 2, with all other patients assumed as PG- SGA A (well nourished), whereas GLIM criteria were assessed for all patients. Finally, the PG-SGA is a subjective diagnostic tool, in contrast, GLIM is an objective measurement which determines patients with malnutrition using objective data and cut-offs, and not by clinician's or patient's judgement [22]. This means that the skill and experience of clinicians may impact less on validity and reliability of the assessment when using the GLIM. For these reasons, studies evaluating the validity of a malnutrition tool, should use an objective reference tool rather than a subjective tool. The GLIM initiative targets the priority to adopt global consensus criteria so that malnutrition prevalence interventions and outcomes can be compared throughout the world [22]. The consensus criteria targets adults in clinical settings, although not specifically for a disease state. The GLIM criteria do not entail the robust detail of comprehensive nutrition assessment but provide a malnutrition diagnosis that may be complemented by more comprehensive assessment, for example the PG-SGA, to provide the basis for individualised treatment plans.

Handgrip strength is a non-invasive, quick and easy method to assess peripheral muscle strength and has been proposed as an indicator of muscle wasting and malnutrition [36]. In cancer patients, muscle wasting is particularly frequent and is associated with increased morbidity and mortality [37]. In fact, in a study in patients with advanced-stage cancer, 6-month mortality was higher amongst malnourished as assessed by the SGA (Subjective Global Assessment) and GLIM criteria using HGS (using 5th percentile cut-off) [38]. It should be kept in mind that hand grip strength is an indicator of upper limb strength and muscle function only and despite its apparent predictive potential, it cannot replace assessment of muscle mass by validated body composition techniques (dual energy x-ray absorptiometry (DXA), bioelectrical impedance analysis (BIA), computed tomography (CT), magnetic resonance imaging (MRI)). Muscle strength tools can be used as a supporting proxy [22]. According to our findings, the addition of a handgrip strength test did not improve the recognition of malnutrition risk when combined with PG-SGA SF. In fact, the sensitivity decreased to 21% (specificity 96%), and the agreement with PG-SGA was poor. Furthermore, the addition of a handgrip strength test also did not improve the recognition of malnutrition when combined to the GLIM criteria.

There is well-established evidence of the PG-SGA's ability to predict clinical outcomes, such as patient survival, postoperative complications, length of stay, quality of life and hospitalisation costs [14]. Given the relatively new nature of the GLIM criteria, there are few studies investigating the predictive ability in a cancer population. Recent studies found that malnutrition according to GLIM criteria with the addition of hand grip strength in hospitalised cancer patients is associated with higher mortality, specifically a 2–3-fold increase in mortality [38,39]. These findings are similar to our results, which confirm that the GLIM criteria are sensitive to identify mortality risk. We note however, that it is difficult to compare findings due to different populations, stages of disease and settings (ambulatory vs. hospitalised), as well as differing hand grip strength references. Thus, more research is necessary to investigate this association.

4.1. Limitations

The principal limitation of our study was removal of the activity box (box 4) from the PG-SGA SF and therefore not using the tool as it was designed. This decision was based on findings by Abbott et al., [16] showing that PG-SGA and additive score combinations of the first three boxes (removing box 4) had higher sensitivity than the MST in ambulatory cancer patients.

Another major limitation of our study was that all patients with MST <2 were assumed as well- nourished PG-SGA A. As this study was initially completed as part of the annual malnutrition prevalence auditing, the usual procedure was to progress to PG-SGA only if MST ≥2. To enable analysis of data, all patients with MST <2 were then assumed as PG-SGA A. This may have incorrectly labelled patients as well-nourished when in fact they could have been malnourished.

Additionally, the study was completed over two separate weeks in two separate years with slightly different methods applied. The addition of patient experience questions was only included in the second year which meant that this valuable information was only obtained from 40% of the study population. Other limitations include our sample size of 246 patients, and heterogeneity of diagnoses.

4.2. Implications for Practice and Research

It is challenging to implement routine malnutrition screening in ambulatory cancer care settings. Resources tend to be stretched, therefore screening tools should be quick and simple. We found that the PG-SGA SF is a valid and suitable screening tool for cancer patients and is a good alternative to the MST. The addition of a handgrip strength test did not add value to screening or diagnosing malnutrition; it created extra work for clinicians and burden to patients and therefore has not been implemented as standard care in our cancer centre. The benefit of the PG-SGA SF in being patient-led aligns with core patient centred concepts and with engaging patients in the care process, and our data showed a good acceptance and patient experience. PG-SGA SF and the GLIM criteria both predict patient mortality, however, there needs to be more large-scale studies to validate the GLIM criteria and compare to reference assessment tools, and body composition techniques.

5. Conclusions

The PG-SGA SF is an accurate, sensitive and specific malnutrition screening tool in the ambulatory cancer care setting. It is patient completed which aligns with patient-centred concepts, and it is well accepted by patients. The addition of a handgrip strength test did not add value, in fact, it reduced the sensitivity in screening or diagnosing malnutrition. The GLIM criteria diagnosed a greater percentage of malnutrition when compared to PG-SGA and the agreement of GLIM to this reference tool is fair. The GLIM criteria should not replace the use of a comprehensive nutrition assessment and should be used in parallel with established and validated tools such as PG-SGA.

Author Contributions: Conceptualization, L.M.D.G., A.A. and B.S.v.d.M.; methodology, L.M.D.G., A.A. and B.S.v.d.M.; validation, L.M.D.G., G.L. and B.S.v.d.M.; formal analysis, L.M.D.G., G.L. and B.S.v.d.M.; writing—original draft preparation, L.D.D.G., G.L. and B.S.v.d.M.; writing—review and editing, L.M.D.G., G.L., and B.S.v.d.M.; supervision, B.S.v.d.M. All authors have read and agreed to the published version of the manuscript.

Funding: This research received no external funding.

Acknowledgments: The authors thank the staff of Mater Hospital for their support and the Bond University Nutrition and Dietetic students for assisting in data collection.

Conflicts of Interest: The authors declare no conflicts of interest.

References

1. Kumar, A.; Saxena, M.; Hammond, N.; Taylor, C.; Thompson, K.; Grattan, S. Near Two-fold Rise in ICD-coded Sepsis-related Hospital Admissions in Australia: An Australian Institute of Health and Welfare Database Analysis (2011–2016). *Aust. Crit. Care* **2019**, *32*, S7. [CrossRef]
2. Hébuterne, X.; Lemarié, E.; Michallet, M.; De Montreuil, C.B.; Schneider, S.M.; Goldwasser, F. Prevalence of Malnutrition and Current Use of Nutrition Support in Patients with Cancer. *J. Parenter. Enter. Nutr.* **2014**, *38*, 196–204. [CrossRef]
3. Arends, J.; Baracos, V.; Bertz, H.; Bozzetti, F.; Calder, P.C.; Deutz, N.E.P.; Erickson, N.; Laviano, A.; Lisanti, M.; Lobo, D.N.; et al. ESPEN expert group recommendations for action against cancer-related malnutrition. *Clin. Nutr.* **2017**, *36*, 1187–1196. [CrossRef]
4. Andreyev, H.J.N.; Norman, A.R.; Oates, J.; Cunningham, D. Why do patients with weight loss have a worse outcome when undergoing chemotherapy for gastrointestinal malignancies? *Eur. J. Cancer* **1998**, *34*, 503–509. [CrossRef]
5. Persson, C.; Glimelius, B. The relevance of weight loss for survival and quality of life in patients with advanced gastrointestinal cancer treated with palliative chemotherapy. *Anticancer. Res.* **2003**, *22*, 3661–3668.
6. Lim, S.L.; Ong, K.C.B.; Chan, Y.H.; Loke, W.C.; Ferguson, M.; Daniels, L. Malnutrition and its impact on cost of hospitalization, length of stay, readmission and 3-year mortality. *Clin. Nutr.* **2012**, *31*, 345–350. [CrossRef]
7. DeWys, W.D.; Begg, C.; Lavin, P.T.; Band, P.R.; Bennett, J.M.; Bertino, J.R.; Cohen, M.H.; Douglass, H.O.; Engstrom, P.F.; Ezdinli, E.Z.; et al. Prognostic effect of weight loss prior tochemotherapy in cancer patients. *Am. J. Med.* **1980**, *69*, 491–497. [CrossRef]
8. Datema, F.R.; Ferrier, M.B.; De Jong, R.J.B. Impact of severe malnutrition on short-term mortality and overall survival in head and neck cancer. *Oral Oncol.* **2011**, *47*, 910–914. [CrossRef]
9. Berg, M.G.A.V.D.; Rasmussen-Conrad, E.L.; Wei, K.H.; Lintz-Luidens, H.; Kaanders, J.H.A.M.; Merkx, M. Comparison of the effect of individual dietary counselling and of standard nutritional care on weight loss in patients with head and neck cancer undergoing radiotherapy. *Br. J. Nutr.* **2010**, *104*, 872–877. [CrossRef]

10. Schmidt, K.N.; Olson, K.; Kubrak, C.; Parliament, M.; Ghosh, S. Validation of the Head and Neck Patient Symptom Checklist as a nutrition impact symptom assessment tool for head and neck cancer patients. *Support. Care Cancer* **2012**, *21*, 27–34. [CrossRef]
11. Um, M.H.; Choi, M.Y.; Lee, S.M.; Lee, I.-J.; Lee, C.G.; Park, Y.K. Intensive nutritional counseling improves PG-SGA scores and nutritional symptoms during and after radiotherapy in Korean cancer patients. *Support. Care Cancer* **2014**, *22*, 2997–3005. [CrossRef]
12. Arends, J.; Bachmann, P.; Baracos, V.; Barthelemy, N.; Bertz, H.; Bozzetti, F.; Fearon, K.; Hütterer, E.; Isenring, E.; Kaasa, S.; et al. ESPEN guidelines on nutrition in cancer patients. *Clin. Nutr.* **2017**, *36*, 11–48. [CrossRef]
13. Kubrak, C.; Jensen, L. Critical Evaluation of Nutrition Screening Tools Recommended for Oncology Patients. *Cancer Nurs.* **2007**, *30*, E1–E6. [CrossRef]
14. Jager-Wittenaar, H.; Ottery, F.D. Assessing nutritional status in cancer. *Curr. Opin. Clin. Nutr. Metab. Care* **2017**, *20*, 322–329. [CrossRef]
15. Gabrielson, D.K.; Scaffidi, D.; Leung, E.; Stoyanoff, L.; Robinson, J.; Nisenbaum, R.; Brezden-Masley, C.; Darling, P.B. Use of an Abridged Scored Patient-Generated Subjective Global Assessment (abPG-SGA) as a Nutritional Screening Tool for Cancer Patients in an Outpatient Setting. *Nutr. Cancer* **2013**, *65*, 234–239. [CrossRef]
16. Abbott, J.; Teleni, L.; McKavanagh, D.; Watson, J.; McCarthy, A.L.; Isenring, E. Patient-Generated Subjective Global Assessment Short Form (PG-SGA SF) is a valid screening tool in chemotherapy outpatients. *Support. Care Cancer* **2016**, *24*, 3883–3887. [CrossRef]
17. Ottery, F.D.; PG-SGA©. PG-SGA/Pt-Global Platform. Available online: http://pt-global.org/?page_id=13. (accessed on 8 July 2020).
18. Di Bella, A.; Blake, C.; Young, A.; Pelecanos, A.; Brown, T. Reliability of Patient-Led Screening with the Malnutrition Screening Tool: Agreement between Patient and Health Care Professional Scores in the Cancer Care Ambulatory Setting. *J. Acad. Nutr. Diet.* **2018**, *118*, 1065–1071. [CrossRef]
19. Norman, K.; Stobäus, N.; Smoliner, C.; Zocher, D.; Scheufele, R.; Valentini, L.; Lochs, H.; Pirlich, M. Determinants of hand grip strength, knee extension strength and functional status in cancer patients. *Clin. Nutr.* **2010**, *29*, 586–591. [CrossRef]
20. Versteeg, K.S.; Blauwhoff-Buskermolen, S.; Buffart, L.; De Van Der Schueren, M.A.; Langius, J.A.E.; Verheul, H.M.W.; Maier, A.B.; Konings, I.R. Higher Muscle Strength Is Associated with Prolonged Survival in Older Patients with Advanced Cancer. *Oncologist* **2017**, *23*, 580–585. [CrossRef]
21. Norman, K.; Wirth, R.; Neubauer, M.; Eckardt, R.; Stobäus, N. The Bioimpedance Phase Angle Predicts Low Muscle Strength, Impaired Quality of Life, and Increased Mortality in Old Patients with Cancer. *J. Am. Med. Dir. Assoc.* **2015**, *16*, 173.e17–173.e22. [CrossRef]
22. Cederholm, T.; Jensen, G.; Correia, M.; Gonzalez, M.; Fukushima, R.; Higashiguchi, T.; Baptista, G.; Barazzoni, R.; Blaauw, R.; Coats, A.; et al. GLIM criteria for the diagnosis of malnutrition—A consensus report from the global clinical nutrition community. *Clin. Nutr.* **2019**, *38*, 1–9. [CrossRef]
23. Isenring, E.; Cross, G.; Daniels, L.; Kellett, E.; Koczwara, B. Validity of the malnutrition screening tool as an effective predictor of nutritional risk in oncology outpatients receiving chemotherapy. *Support. Care Cancer* **2006**, *14*, 1152–1156. [CrossRef]
24. Dodds, R.M.; Syddall, H.E.; Cooper, R.; Benzeval, M.; Deary, I.J.; Dennison, E.M.; Der, G.; Gale, C.R.; Inskip, H.; Jagger, C.; et al. Grip Strength across the Life Course: Normative Data from Twelve British Studies. *PLoS ONE* **2014**, *9*, e113637. [CrossRef]
25. Planas, M.; Álvarez-Hernández, J.; León-Sanz, M.; Celaya-Pérez, S.; Araujo, K.; De Lorenzo, A.G.; Researchers on behalf of the PREDyCES®. Prevalence of hospital malnutrition in cancer patients: A sub-analysis of the PREDyCES®study. *Support. Care Cancer* **2016**, *24*, 429–435. [CrossRef]
26. Onesti, J.K.; Guttridge, D.C. Inflammation Based Regulation of Cancer Cachexia. *BioMed Res. Int.* **2014**, *2014*, 1–7. [CrossRef]
27. Raslan, M.; Gonzalez, M.C.; Dias, M.C.G.; Nascimento, M.; Castro, M.; Marques, P.; Segatto, S.; Torrinhas, R.S.; Cecconello, I.; Waitzberg, D.L. Comparison of nutritional risk screening tools for predicting clinical outcomes in hospitalized patients. *Nutrition* **2010**, *26*, 721–726. [CrossRef]
28. Viera, A.J.; Garrett, J.M. Understanding interobserver agreement: The kappa statistic. *Fam. Med.* **2005**, *37*, 360–363.

29. McHugh, M.L. Interrater reliability: The kappa statistic. *Biochem. Medica* **2012**, 276–282. [CrossRef]
30. Campbell, K.L.; Bauer, J.D.; Ikehiro, A.; Johnson, D.W. Role of Nutrition Impact Symptoms in Predicting Nutritional Status and Clinical Outcome in Hemodialysis Patients: A Potential Screening Tool. *J. Ren. Nutr.* **2013**, *23*, 302–307. [CrossRef]
31. Mendonsa, R.D.; Appaya, P. Psychiatric morbidity in outpatients of gynecological oncology clinic in a tertiary care hospital. *Indian J. Psychiatry* **2010**, *52*, 327–332. [CrossRef]
32. Nho, J.-H.; Kim, S.R.; Kwon, Y.S. Depression and appetite: Predictors of malnutrition in gynecologic cancer. *Support. Care Cancer* **2014**, *22*, 3081–3088. [CrossRef]
33. Shaw, C.; Fleuret, C.; Pickard, J.M.; Mohammed, K.; Black, G.; Wedlake, L. Comparison of a novel, simple nutrition screening tool for adult oncology inpatients and the Malnutrition Screening Tool (MST) against the Patient-Generated Subjective Global Assessment (PG-SGA). *Support. Care Cancer* **2014**, *23*, 47–54. [CrossRef]
34. Bauer, J.D.; Capra, S.M.; Ferguson, M. Use of the scored Patient-Generated Subjective Global Assessment (PG-SGA) as a nutrition assessment tool in patients with cancer. *Eur. J. Clin. Nutr.* **2002**, *56*, 779–785. [CrossRef]
35. Read, J.A.; Crockett, N.; Volker, D.H.; MacLennan, P.; Choy, S.T.B.; Beale, P.; Clarke, S.J. Nutritional Assessment in Cancer: Comparing the Mini-Nutritional Assessment (MNA) With the Scored Patient-Generated Subjective Global Assessment (PGSGA). *Nutr. Cancer* **2005**, *53*, 51–56. [CrossRef]
36. Norman, K.; Schütz, T.; Kemps, M.; Lübke, H.J.; Lochs, H.; Pirlich, M. The Subjective Global Assessment reliably identifies malnutrition-related muscle dysfunction. *Clin. Nutr.* **2005**, *24*, 143–150. [CrossRef] [PubMed]
37. Ryan, A.M.; Power, D.G.; Daly, L.E.; Cushen, S.J.; Bhuachalla Ēadaoin, N.; Prado, C.M. Cancer-associated malnutrition, cachexia and sarcopenia: The skeleton in the hospital closet 40 years later. *Proc. Nutr. Soc.* **2016**, *75*, 199–211. [CrossRef] [PubMed]
38. Contreras-Bolívar, V.; Sánchez-Torralvo, F.J.; Ruiz-Vico, M.; González-Almendros, I.; Barrios, M.; Padín, S.; Alba, E.; Olveira-Fuster, A.G.; Bolívar, C.; Torralvo, S.; et al. GLIM Criteria Using Hand Grip Strength Adequately Predict Six-Month Mortality in Cancer Inpatients. *Nutrients* **2019**, *11*, 2043. [CrossRef] [PubMed]
39. Yilmaz, M.; Atilla, F.D.; Sahin, F.; Saydam, G. The effect of malnutrition on mortality in hospitalized patients with hematologic malignancy. *Support. Care Cancer* **2019**, *28*, 1441–1448. [CrossRef]

nutrients
MDPI

Article

Oral Nutritional Supplementation Affects the Dietary Intake and Body Weight of Head and Neck Cancer Patients during (Chemo) Radiotherapy

Isabela Borges Ferreira [1], Emanuelle do Nascimento Santos Lima [1], Paula Philbert Lajolo Canto [2], Cristiana Araújo Gontijo [1], Yara Cristina de Paiva Maia [1] and Geórgia das Graças Pena [1,*]

1 Graduate Program in Health Sciences, School of Medicine, Federal University of Uberlandia, 1720 Para Avenue, 2H, Uberlandia MG 38400-902, Brazil; isaborgesferreira@gmail.com (I.B.F.); emanuellensl@yahoo.com.br (E.d.N.S.L.); cristianaagontijo@hotmail.com (C.A.G.); yara.maia@ufu.br (Y.C.d.P.M.)
2 Department of Oncology, Clinical Hospital of Federal University of Uberlandia, 1888 Para Avenue, Uberlandia MG 38405-320, Brazil; paula.lajolo@oncocentro.com.br
* Correspondence: georgia@ufu.br; Tel.: +55-34-3225-8584

Received: 8 July 2020; Accepted: 15 August 2020; Published: 20 August 2020

Abstract: Considering the symptoms of (chemo) radiotherapy and the reduction in food intake in head and neck cancer (HNC) patients, this study aimed to investigate the association between treatment time points and oral nutritional supplementation (ONS) on dietary intake to estimate the frequency of energy and nutrient inadequacy, and also to evaluate body weight changes (BWC). Dietary intake data of 65 patients were obtained from 24-h dietary recalls and prevalence of inadequacy was calculated before or at the beginning (T0), in the middle (T1), and at the end of treatment (T2). BWC were calculated as the weight difference considering the previous weight reported and/or measured. Energy and macronutrient intake decreased in T1 and then improved in T2 ($p < 0.001$ for both). Micronutrient intake increased during treatment due to ONS use, but still presented a high probability of inadequate intake. In particular, calcium, magnesium, and vitamin B6 showed almost 100% of probability of inadequacy for those who did not use ONS. Finally, overweight patients suffered a higher weight accumulated deficit with a delta of −15 kg compared to other BMI (body mass index) categories. Therefore, we strongly recommend initiating nutritional counseling in conjunction with prophylactic ONS prescription from diagnosis to adjust nutrient intake and minimize weight loss.

Keywords: food consumption; weight loss; malnutrition; dietary supplements; head and neck neoplasms

1. Introduction

Head and neck cancer (HNC) are the seventh most common malignant tumors in the world [1]. Among the modalities of treatment, chemoradiotherapy (CRT) is related to symptoms such as oral mucositis, xerostomia, and dysgeusia, which could affect dietary intake [2,3]. Aside from symptoms resulting from treatment, these patients also have other negative symptoms as the tumor can cause problems when chewing and swallowing [4], and make eating difficult and painful [5].

These symptoms can cause a reduction in dietary intake [5], and are associated with weight loss and malnutrition [6,7], worse quality of life [4], infection, higher hospital readmissions, longer length of hospital stay, and mortality [8]. As a result, a marked change in the consistency of food consumption by HNC patients has been observed during the treatment period that may interfere with energy adequacy [5,9] and lead to reduced macronutrient intake. Furthermore, inadequate intake of micronutrients such as vitamins D, E, C, folate, calcium, iron, and magnesium has been described in HNC patients, and oral nutritional supplementation (ONS) has been required in order to achieve recommended levels [10,11], since micronutrients are important for enzymatic reactions that impact the metabolism as a whole [12].

According to the United Kingdom National Multidisciplinary Guidelines [13], nutritional support is an important part of the treatment of HNC patients. Nutritional intervention during treatment is indicated in order to prevent weight loss, increase food intake, and decrease treatment interruptions. Thus, when oral nutrition is inadequate, ONS use and tube feeding are indicated. HNC patients who receive nutritional counseling and use nutritional support show improvement in weight loss, quality of life and survival [14–17].

Thus, nutritional support of HNC patients including ONS use is important. However, the levels of both dietary intake and ONS contribution to macro- and micronutrient consumption during treatment are currently unknown or understudied. Studies in the literature assess the presence of individual nutritional counseling [18] and use or not of ONS [19] with outcomes such as weight [15,20], quality of life, mortality, and nutritional status [21]. However, as these variables are abstract, studies that account for or assist in quantifying the nutrient intake are necessary, since the quantification of consumption is often neglected in clinical practice and little explored in studies due to the complexity of analysis. Thus, with this evaluation, it is possible to obtain an assertive nutritional approach. Moreover, most studies evaluating dietary intake in HNC patients are long-term [10,22], and do not assess the effects and acute treatment changes presented by these patients.

To the best of our knowledge, this is the first prospective study to evaluate macro and micronutrient intake during treatment; that is, in the short-term. This is important because it allows dietary nutritional monitoring to be frequently carried out in order to minimize the possible cumulative losses that the patient has since the onset of symptoms and which are often not properly valued in these studies. In addition, there are no data in the literature on the prevalence of inadequate dietary intake in these patients. It is therefore important to identify the treatment time point at which the impact on dietary intake is greatest, so that an early assessment of nutritional changes can be made and the above-mentioned negative impacts minimized through nutritional counseling. Furthermore, it is possible to minimize weight loss by considering not only nutritional counseling, but the identification of specific needs, taking care to not neglect patients who do not demonstrate physical malnutrition so clearly.

We hypothesized that macro- and micronutrient intake was reduced in HNC patients during treatment. Thus, the aim of this prospective study was to investigate the association between treatment time points and ONS on dietary intake to estimate the frequency of energy and nutrient inadequacy as well as to evaluate body weight changes (BWC) in HNC patients during (chemo) radiotherapy.

2. Materials and Methods

2.1. Study Design and Ethical Aspects

A prospective observational study was carried out with HNC patients during (chemo) radiotherapy. These patients were recruited from the outpatient treatment of a tertiary university hospital, which was the regional referral center for HNC patients undergoing antineoplasic treatment in the city of Uberlandia, Minas Gerais, Brazil, between July 2017 and November 2018. The patients were evaluated at three time points: before or at the beginning of treatment (T0); in the middle (T1 ~four weeks,) and at the final treatment (T2 ~eight weeks).

This study was approved by the Human Research Ethics Committee (protocol number 65340116.8.0000.5152) and all participants signed a free and informed consent form. The entire study was conducted based on the standards of the Declaration of Helsinki [23].

2.2. Sample Size and Eligibility Criteria

In order to estimate the sample size required for this study, we used G* Power software, version 3.1 (Heinrich-Heine-University Düsseldorf, Düsseldorf, Germany). Considering a single group of individuals and three measurements, the sample size calculations were based on an F-test repeated-measures ANOVA with an effect size of 0.25, an α level of 0.05, and 95% power. The result of the calculation required a minimum sample size of forty-five patients. So, considering a 20% adjustment for possible losses, a minimum of fifty-four patients was needed at baseline (T0).

The inclusion criteria were patients aged 18 years or older, diagnosed with primary malignant tumors in the head and neck region who were undergoing radiotherapy (RT), chemotherapy (CT), or a combination of these modalities, with or without surgery, independent of tumor stage, and were invited to participate. Patients were considered to be T0 when treatment had not been initiated or at the beginning of treatment, before presenting collateral effects. Patients with metastasis at T0 or who had been previously treated with RT and/or CT for other types of cancer in the last 10 years were excluded.

During the study period, 140 patients were approached, 25 declined to participate, and 24 did not meet the inclusion criteria, resulting in 91 patients (60 patients with complete dietary intake data in T0). Of the 91, two died, 16 lost follow-up, and three refused to continue in the study, totaling 70 patients in T1 (56 patients with complete dietary intake data in T1). Considering the total participants in T1, one died and six lost follow-up, totaling 63 patients in T2 (53 patients with complete dietary intake data in T2). With respect to patients with complete intake data, five patients were included only at T1 and T2 and nine patients were included only at T0 and T2, resulting in 65 patients with a complete dietary assessment.

2.3. Demographic, Clinical, and Nutritional Assessment

Characteristics such as age (years), sex (female/male), clinical diagnosis, tumor site, tumor stage by American Joint Committee on Cancer–AJCC [24], treatment schedule, ONS or gastric or enteral tubes use and nutritional counseling were obtained from an initial structured questionnaire and medical records.

Height and body weight were measured at three study time points using standard protocols [25] and weight at six (habitual weight) and one month ago was self-reported. Body Mass Index (BMI) was calculated as body weight (kg)/height (m^2) for assessment and classification of nutritional status. The patients were classified into three groups based on the World Health Organization criteria of underweight (<18.5 kg/m^2), normal weight (18.5–24.9 kg/m^2), and overweight (≥25.0 kg/m^2) [25].

BWC were calculated as the weight difference considering the weight for six months (habitual weight) and one month self-reported before treatment, and measured at the three time points of the study (T0, T1, and T2). BWC calculation considering five points (two points before treatment and the three treatment time points): BWC T_{6mo} = 0; BWC T_{1mo} = weight for 1 month − habitual weight; BWC T0 = weight T0 − habitual weight; BWC T1 = weight T1 − habitual weight; BWC T2 = weight T2 − habitual weight. To calculate the habitual BMI, weight for six months (habitual weight) was used and height was measured at T0. BWC calculation considering three points (the three treatment time points): BWC T0 = 0; BWC T1 = weight T1 − weight T0; BWC T2 = weight T2 − weight T0. To calculate the BMI at T0, the weight and height measured at T0 were used.

2.4. Dietary Assessment

Dietary and ONS intake data were obtained from 24-h dietary recalls (24HR) applied at three study time point (T0, T1, and T2). At each time point, the first 24HR was applied in person and other two by telephone interview on non-consecutive days including weekend days, in order to record the eating habits of the study participants more precisely, ideally totaling nine 24HRs for each patient.

Data were collected using the United States Department of Agriculture Multiple-Pass Method, which guides the respondent to respond to 24HR in five steps [26]. ONS intake was calculated using the manufacturers' labels in order to include all nutrients coming from this source. The same process was used for patients who were given nutritional therapy by tube feeding. Since the patients can use different brands of the ONS, depending on the chosen manufacturer, the quantity and brand labels were properly registered in order to provide complete accounting of nutrient supply.

The following nutrients were analyzed using 24HR: total energy (kcal), carbohydrate (g), protein (g), lipid (g), dietary fiber (g), monounsaturated, polyunsaturated and saturated fats (g), total cholesterol (mg), thiamine (mg), riboflavin (mg), niacin (mg), vitamin B6 (mg), vitamin C (mg), iron (mg), magnesium (mg), zinc (mg), calcium (mg), phosphorus (mg), manganese (mg), potassium (mg), and sodium (mg). Nutrient content was estimated by Dietpro software, version 5.8.1 (Dietpro Viçosa, MG, BR), using, for preference, the Brazilian Table of Food Composition [27]. For foods not found in this table, the United States Department of Agriculture [28] table was used. Nutritional content from food or supplements not found in the software's tables were added based on their labels.

Energy and nutrient values were adjusted for intra-individual variability due to intrinsic dietary intake variability, in order to obtain an individual consumption estimate of energy and nutrients using the Personal Computer version of the Software for Intake Distribution Estimation (PC-SIDE) (Department of Statistics, Iowa State University, Iowa, USA), following the methodology described by Nusser et al. [29]. Subsequently, these were adjusted by the residual method for the sample total energy in order to adjust nutrient estimates [30].

Prevalence of inadequacy was calculated by the Estimated Average Requirement (EAR) method as a cut-off point [31]. The Z value was calculated ((EAR—average intake)/standard deviation)) and the Z table curve was consulted to verify the corresponding percentage of individuals with intakes below EAR. For this evaluation, the values of energy and nutrients adjusted only for intra-individual variability were used.

For fiber, manganese, potassium, and sodium, for which there are no established EAR values, an intake comparison was made with their respective adequate intake (AI) values. When these nutrients showed an intake above AI, adequacy regarding tolerable upper intake level (UL) was verified. Macronutrient distribution to total energy value was analyzed using acceptable macronutrients distribution range (AMDR) values as a reference [32].

For energy (25 kcal/kg/day) and protein intake (1 g/kg/day), the European Society for Clinical Nutrition and Metabolism (ESPEN) recommendations were used as reference values [33] to understand whether individuals are capable of achieving the recommended minimum levels of intake of these macronutrients. For cholesterol and monounsaturated, polyunsaturated, and saturated fats, we used the recommendations of the Food and Agriculture Organization of the United Nations [34].

2.5. Statistical Analyses

Variable distributions were evaluated by the Kolmogorov–Smirnov test. Descriptive statistics were shown in percentage, mean, and standard deviation to describe the characteristics of the investigated population.

Generalized estimating equations (GEE) models were used to determine the association of treatment time points (T0, T1, and T2), ONS use and treatment time points, and ONS interactions (independent variables) with nutrient consumption (dependent variables). GEE is a method that considers the association between different observations in the same individual in prospective studies, performing a better evaluation of repeated-measures data [35]. The gamma, linear, or Tweedie distribution models were individually tested for all outcomes. Lower quasi-likelihood under the independence model criterion (QIC) was observed in the gamma with the log-link model, and was chosen for GEE analysis. The Bonferroni post-hoc test was used to adjust for multiple comparisons. Type of treatment, sex, age (years), tumor site, and stage were considered as confounders.

ONS use was grouped in order to be evaluated as an exposure in the GEE models. Since the ONS can be indicated at the beginning, middle, or final treatment, the individuals were categorized according the frequency of ONS use by time points: individuals who used ONS 2/3 times (higher frequency of use) and those who used ONS 0/1 time (no or lower frequency of use). This strategy was also used to fix this exposure to analyze the longitudinal effect of ONS use on the dietary intake. For statistical tests not performed by GEE, the individuals were categorized into "with ONS" or "without ONS" at each treatment time point. Confidence interval (CI) of 95% and p-value < 0.05 were considered as levels of statistical significance. All data were analyzed using Statistical Package for Social Sciences (SPSS), version 25.0 (SPSS Inc., Chicago, IL, USA).

3. Results

Of the 65 patients, the mean age was 59.8 ± 10.1 years, 53 (81.5%) patients were male, 26 (40.0%) had a tumor in the larynx, followed by oral cavity 21 (32.3%), and pharynx 14 (21.5%). The majority of patients were at an advanced (T3–T4; 58.4%) stage of cancer. Among the treatment types, most prevalent were CRT 33 (50.8%) and RT 16 (24.6%) (Table 1). Antineoplasic treatment lasted approximately eight weeks with a daily RT session from Monday to Friday, totaling 38 to 40 sessions. According to the institutional protocol, patients underwent RT with a total final radiation dose of 70 or 72 Gy with daily doses of 180 or 200 cGy. Mean ± standard deviation of the number of RT sessions performed by the study patients was 1.58 ± 2.3 at time T0, 20.27 ± 3.7 at time T1, and 36.27 ± 3.7 at time T2. For patients undergoing CRT, the protocol consisted of weekly cisplatin during the radiotherapy course.

Table 1. Clinical characteristics of patients.

Variables	Total Sample T0 (n = 91), T1 (n = 70), T2 (n = 65)	Analyzed Sample T0 (n = 60), T1 (n = 56), T2 (n = 53) n (%) or Mean ± SD
Age (years)	60.6 ± 10.9	59.8 ± 10.1
Sex, male	70 (76.9)	53 (81.5)
Underweight patients by BMI		
T0	16 (18.0)	9 (13.8)
T1	16 (23.9)	14 (23.3)
T2	16 (25.0)	13 (22.0)
Normal weight patients by BMI		
T0	43 (48.3)	31 (47.7)
T1	35 (52.2)	30 (50.0)
T2	35 (54.7)	33 (55.9)
Overweight patients by BMI		
T0	30 (33.7)	25 (38.5)
T1	16 (23.9)	16 (26.7)
T2	13 (20.3)	13 (22.0)
Oral nutritional supplements use		
T0	32 (35.2)	18 (30.0)
T1	47 (67.1)	35 (62.5)
T2	47 (72.3)	36 (67.9)
Gastric or enteral tubes		
T0	10 (11.0)	4 (6.1)
T1	14 (20.2)	13 (20.9)
T2	11 (17.2)	10 (17.0)
Nutritional counseling		
T0	20 (22.0)	12 (18.5)
T1	44 (63.8)	38 (61.3)
T2	44 (67.7)	41 (68.3)
Tumor site		
Oral cavity [a]	30 (33.0)	21 (32.3)
Nasal cavity	4 (4.4)	3 (4.6)
Larynx	32 (35.2)	26 (40.0)
Pharynx [b]	22 (24.2)	14 (21.5)
Other [c]	3 (3.3)	1 (1.5)

Table 1. *Cont.*

Variables	Total Sample T0 (*n* = 91), T1 (*n* = 70), T2 (*n* = 65)	Analyzed Sample T0 (*n* = 60), T1 (*n* = 56), T2 (*n* = 53) *n* (%) or Mean ± SD
T Stage		
T1	8 (8.8)	8 (12.2)
T2	21 (23.1)	17 (26.1)
T3	28 (30.8)	21 (32.3)
T4	28 (30.8)	17 (26.1)
TX	4 (4.4)	2 (3.1)
Not specified or unknown	2 (2.2)	
N Stage		
N0	38 (41.8)	30 (46.2)
N1	19 (20.9)	17 (26.2)
N2	18 (19.8)	11 (16.9)
N3	8 (8.8)	4 (6.2)
NX	6 (6.6)	3 (4.6)
Not specified or unknown	2 (2.2)	
M Stage		
M0	55 (60.4)	40 (61.5)
M1	4 (4.4)	2 (3.1)
MX	26 (28.6)	20 (30.8)
Not specified or unknown	6 (6.6)	3 (4.6)
Clinical Stage		
I	8 (8.8)	8 (12.3)
II	13 (14.3)	10 (15.4)
III	24 (26.4)	19 (29.2)
IV	42 (46.2)	28 (43.1)
Not specified or unknown	4 (4.4)	
Mode of treatment		
Radiotherapy	21 (23.9)	16 (24.6)
Surgery and radiotherapy	10 (11.4)	10 (15.4)
Chemoradiotherapy	37 (42.0)	33 (50.8)
Surgery and chemoradiotherapy	7 (8.0)	6 (9.2)
Chemotherapy	1 (1.1)	
Surgery	5 (5.7)	
Other [d]	7 (8.0)	

Abbreviations: BMI, body mass index; SD, standard deviation; Treatment time points: T0, before or at beginning of treatment; T1, middle of treatment; T2, final of treatment. [a] tongue, mouth floor, and lip; [b] hypopharynx, oropharynx, and nasopharynx; [c] jaw, cervical and parathyroid; [d] Loss of follow-up before starting treatment.

Regarding nutritional status performed by BMI, the frequency of underweight increased in the middle 14 (23.3%), and at the end of treatment 13 (22.0%), and overweight decreased in the middle 16 (26.7%) and at the end of treatment 13 (22.0%), both compared to before treatment. ONS use increased during treatment (Table 1).

The association of treatment time points and group of ONS use on energy, macro-, and micronutrient intake is shown in Table 2. In general, there was a significant reduction in energy ($p < 0.001$), macronutrients ($p < 0.001$), and cholesterol ($p < 0.001$) intake from the beginning (T0) to the middle of treatment (T1), with the increase at the end of treatment (T2). In addition, those who used ONS 2/3 times consumed more protein and less polyunsaturated fat (1.32 g/kg/day and 16.95 g, respectively) than those who used ONS 0/1 time (1.18 g/kg/day and 19.32 g, respectively) at all treatment time points, with a difference of +0.14 g/kg/day for protein ($p = 0.031$) and −2.37 g of polyunsaturated fat ($p = 0.005$). For micronutrients, except for niacin and vitamin C, an increase in intake from the beginning (T0) to the middle of treatment (T1) was observed and the values were maintained at the end of treatment (T2). The use of ONS 2/3 times increased intake of all micronutrients, except potassium and sodium (Table 2).

Table 2. Association of treatment time points and oral nutritional supplementation on energy, macro-, and micronutrient intake in head and neck cancer patients during (chemo) radiotherapy.

Dependent Variables Mean SE	Independent Variables—Treatment Time Points, Oral Nutritional Supplementation, and Treatment Time Points with Oral Nutritional Supplementation													
	T0 (n = 60)	T1 (n = 56)	T2 (n = 53)	p-Value	ONS 0/1 Time Overall (n = 82)	ONS 2/3 Times Overall (n = 87)	p-Value	T0 ONS 0/1 Time (n = 32)	T0 ONS 2/3 Times (n = 28)	T1 ONS 0/1 Time (n = 26)	T1 ONS 2/3 Times (n = 30)	T2 ONS 0/1 Time (n = 24)	T2 ONS 2/3 Times (n = 29)	p-Value
Energy (kcal)	1829.55 [a] 85.12	1511.03 [b] 66.80	1804.08 [a] 82.33	**<0.001**	1707.20 73.31	1709.87 87.62	**0.976**	1859.03 100.05	1800.54 117.43	1505.06 81.79	1517.03 91.21	1778.34 96.57	1830.19 98.55	0.715
Energy (kcal/kg/day) [d]	27.62 [a] 1.41	23.48 [b] 1.47	29.44 [a] 1.52	**<0.001**	25.97 1.30	27.51 1.93	0.422	27.25 1.47	28.00 2.21	22.61 1.73	24.38 1.97	28.42 1.72	30.49 2.20	0.829
Carbohydrate (g)	230.47 [a] 3.86	204.12 [b] 3.10	246.05 [c] 3.49	**<0.001**	224.81 3.44	227.62 3.51	0.553	226.18 4.98	234.83 5.69	205.89 4.27	202.37 4.00	243.97 5.34	248.15 4.80	0.296
Protein (g)	88.41 [a] 1.61	70.43 [b] 1.27	80.08 [c] 1.40	**<0.001**	77.80 1.25	80.83 1.59	0.123	88.42 2.14	88.41 2.42	69.36 1.78	71.51 1.68	76.79 1.77	83.52 2.20	0.147
Protein (g/kg/day) [d]	1.34 [a] 0.06	1.10 [b] 0.05	1.31 [a] 0.06	**<0.001**	1.18 [a] 0.05	1.32 [b] 0.07	**0.031**	1.29 0.06	1.39 0.09	1.04 0.05	1.18 0.06	1.22 0.06	1.42 0.08	0.346
Lipids (g)	77.23 [a] 1.07	60.42 [b] 1.00	68.31 [c] 1.26	**<0.001**	68.88 1.08	67.75 1.18	0.498	79.23 1.40	75.28 1.62	59.89 1.35	60.96 1.37	68.86 2.01	67.76 1.78	0.176
Monounsaturated fat (g)	20.89 [a] 0.57	15.41 [b] 0.38	17.75 [c] 0.59	**<0.001**	18.37 0.49	17.40 0.54	0.169	21.72 0.71	20.10 0.76	15.47 0.54	15.36 0.57	18.45 0.78	17.07 0.85	0.417
Polyunsaturated fat (g)	21.33 [a] 0.86	15.43 [b] 0.54	18.02 [c] 0.70	**<0.001**	19.32 [a] 0.72	16.95 [b] 0.69	**0.005**	23.27 1.14	19.54 1.02	16.05 0.69	14.83 0.73	19.32 0.95	16.81 0.92	0.440
Saturated fat (g)	22.50 [a] 0.49	19.15 [b] 0.41	21.69 [a] 0.59	**<0.001**	20.44 0.43	21.71 0.52	0.061	22.43 0.58	22.57 0.71	18.52 0.56	19.80 0.61	20.55 0.95	22.90 0.75	0.178
Cholesterol (mg)	289.81 [a] 14.89	196.10 [b] 11.29	263.94 [a] 16.08	**<0.001**	247.68 12.59	245.57 15.85	0.906	291.92 17.82	287.72 21.51	196.82 14.40	195.37 14.51	264.43 19.31	263.45 23.98	0.995
Calcium (mg)	489.32 [a] 24.57	771.70 [b] 32.74	812.89 [b] 52.45	**<0.001**	562.76 [a] 25.13	808.59 [b] 50.56	**<0.001**	398.89 26.65	600.26 48.39	653.39 32.32	911.43 57.70	683.83 61.90	966.30 78.40	0.759
Iron (mg)	7.76 [a] 0.43	9.98 [b] 0.46	10.19 [b] 0.73	**<0.001**	8.33 [a] 0.32	10.25 [b] 0.85	**0.019**	7.46 0.40	8.07 0.71	8.75 0.51	11.39 1.00	8.87 0.72	11.71 1.29	0.148
Fiber (g)	20.31 [a] 0.82	17.86 [b] 0.86	19.80 [a,b] 0.65	**0.032**	20.00 0.75	18.62 0.82	0.174	21.56 1.14	19.14 1.05	18.01 1.17	17.72 1.18	20.60 0.80	19.03 0.96	0.582

Table 2. *Cont.*

Dependent Variables Mean SE	T0 (n = 60)	T1 (n = 56)	T2 (n = 53)	p-Value	ONS 0/1 Time Overall (n = 82)	ONS 2/3 Times Overall (n = 87)	p-Value	T0 ONS 0/1 Time (n = 32)	T0 ONS 2/3 Times (n = 28)	T1 ONS 0/1 Time (n = 26)	T1 ONS 2/3 Times (n = 30)	T2 ONS 0/1 Time (n = 24)	T2 ONS 2/3 Times (n = 29)	p-Value
	Independent Variables—Treatment Time Points, Oral Nutritional Supplementation, and Treatment Time Points with Oral Nutritional Supplementation													
Phosphorus (mg)	1007.34 ^{a} 24.66	1015.58 ^{a} 24.22	1127.24 ^{b} 23.77	**<0.001**	990.57 ^{a} 21.59	1110.16 ^{b} 30.58	**0.001**	960.26 27.12	1056.73 42.30	941.74 27.69	1095.21 34.85	1074.84 35.98	1182.20 33.48	0.373
Magnesium (mg)	212.44 ^{a} 8.74	252.40 ^{b} 9.94	251.03 ^{b} 9.69	**<0.001**	222.03 ^{a} 7.84	254.86 ^{b} 11.20	**0.005**	206.92 9.81	218.10 13.65	223.32 10.14	285.27 16.22	236.85 11.69	266.06 14.00	0.053
Manganese (mg)	1.80 ^{a} 0.08	2.15 ^{b} 0.09	2.42 ^{b} 0.11	**<0.001**	1.97 ^{a} 0.08	2.26 ^{b} 0.10	**0.011**	1.77 ^{a} 0.09	1.84 ^{a,c} 0.12	1.87 ^{a,c} 0.10	2.48 ^{b} 0.14	2.30 ^{b,c} 0.15	2.54 ^{b} 0.16	0.034
Niacin (mg)	15.35 0.57	15.78 0.58	16.50 0.47	0.120	15.03 ^{a} 0.43	16.75 ^{b} 0.63	**0.019**	15.15 0.70	15.55 0.88	14.08 0.79	17.68 0.88	15.92 0.56	17.10 0.72	0.067
Potassium (mg)	3648.17 ^{a,b} 100.35	3446.59 ^{a} 79.65	3718.62 ^{b} 97.27	**0.004**	3531.44 89.78	3675.18 90.79	0.161	3635.20 119.84	3661.19 137.94	3322.53 92.88	3575.28 122.48	3646.36 130.33	3792.31 112.15	0.460
Riboflavin (mg)	0.95 ^{a} 0.04	1.29 ^{b} 0.05	1.37 ^{b} 0.06	**<0.001**	1.01 ^{a} 0.04	1.39 ^{b} 0.07	**<0.001**	0.80 0.05	1.12 0.08	1.10 0.06	1.51 0.08	1.19 0.09	1.58 0.10	0.893
Sodium (mg)	3112.91 ^{a} 79.57	2393.95 ^{b} 68.84	2774.00 ^{c} 77.75	**<0.001**	2825.23 57.12	2666.08 85.65	0.052	3209.89 77.75	3018.87 113.46	2489.29 89.68	2302.26 94.11	2822.25 93.60	2726.58 98.14	0.734
Thiamine (mg)	1.01 ^{a} 0.04	1.21 ^{b} 0.05	1.37 ^{c} 0.06	**<0.001**	1.06 ^{a} 0.04	1.33 ^{b} 0.06	**<0.001**	0.93 0.04	1.09 0.06	1.03 0.06	1.43 0.08	1.25 0.06	1.50 0.09	0.098
Vitamin B6 (mg)	0.59 ^{a} 0.03	0.93 ^{b} 0.05	0.96 ^{b} 0.05	**<0.001**	0.68 ^{a} 0.04	0.96 ^{b} 0.06	**<0.001**	0.51 0.04	0.68 0.06	0.73 0.06	1.19 0.09	0.84 0.06	1.10 0.08	0.126
Vitamin C (mg)	77.85 5.91	96.64 6.54	91.48 7.76	0.054	74.67 ^{a} 4.32	104.40 ^{b} 9.20	**0.001**	63.93 5.61	94.79 11.32	82.27 7.26	113.53 11.04	79.15 9.88	105.73 12.43	0.894
Zinc (mg)	12.50 0.47	11.59 0.42	12.87 0.52	**0.022**	11.50 ^{a} 0.36	13.17 ^{b} 0.62	**0.008**	12.40 ^{a,b} 0.51	12.59 ^{a,b} 0.62	11.03 ^{b} 0.51	12.18 ^{b} 0.68	11.13 ^{b} 0.60	14.89 ^{a} 0.94	0.009

Abbreviations: ONS, oral nutritional supplementation; SE, standard error; Treatment time points: T0, before or at beginning of treatment; T1, middle of treatment; T2, final of treatment; Adjusted for intra-individual variability, proposed by [29]; Adjusted for total energy consumption by residual method, proposed by [30]. Generalized estimating equations model, adjusted: treatment, age, sex, tumor site, and stage. Bonferroni post-hoc test: Different superscript letters represent statistical difference (^{a,b,c}) in pairwise comparisons, $p < 0.05$. Significant tests of model effects shown in bold. Overall ONS is the total patient data values representing the average of the three treatment time points (n = 169). [d] Reference for minimum recommended intake: 25 kcal/kg/day and 1 g protein/kg/day [33].

Percentage and prevalence of energy and nutrient inadequacy is shown in Table 3 for macronutrients and Table 4 for micronutrients. In general, a high percentage and prevalence of inadequacy was observed for energy and nutrient intake, especially in patients who did not use ONS. Macronutrients with the highest percentage of inadequacy according to AMDR were carbohydrates, followed by lipids, monounsaturated fats, and saturated fats. Moreover, even using ONS at T1, protein (grams), lipids, monounsaturated, polyunsaturated, and saturated fat intake presented a high percentage of inadequacy.

Table 3. Percentage of macronutrient inadequacy, protein in g/kg/day, energy in kcal/kg/day, and mean values and standard deviation of cholesterol intake in head and neck cancer patients during (chemo) radiotherapy.

Energy and Nutrients	Percentage of Inadequacy *n* (%)								
	T0			T1			T2		
	Total (*n* = 60)	Without ONS (*n* = 42)	With ONS (*n* = 18)	Total (*n* = 56)	Without ONS (*n* = 21)	With ONS (*n* = 35)	Total (*n* = 53)	Without ONS (*n* = 17)	With ONS (*n* = 36)
Energy (>25 kcal/kg/day) [b]	16 (26.7)	13 (31.0)	3 (16.7)	23 (43.4)	10 (50.0)	13 (39.4)	11 (21.2)	8 (47.1)	3 (8.6)
Carbohydrate (45–65%) [a]	35 (58.3)	26 (61.9)	9 (50.0)	31 (55.4)	13 (61.9)	18 (51.4)	27 (50.9)	10 (58.8)	17 (47.2)
Protein (10–35%) [a]	3 (5.0)	3 (7.1)	0 (0.0)	3 (5.4)	0 (0.0)	3 (8.6)	0 (0.0)	0 (0.0)	0 (0.0)
Protein (>1 g/kg/day) [b]	7 (11.7)	6 (14.3)	1 (5.6)	16 (30.2)	8 (40.0)	8 (24.2)	4 (7.7)	4 (23.5)	0 (0.0)
Lipids- 20–35%) [a]	32 (53.3)	23 (54.8)	9 (50.0)	31 (55.4)	11 (52.4)	20 (57.1)	20 (37.7)	10 (58.8)	10 (27.8)
Monounsaturated fat (15–20%) [c]	57 (95.0)	39 (92.9)	18 (100.0)	56 (100.0)	21 (100.0)	35 (100.0)	52 (98.1)	16 (94.1)	36 (100.0)
Polyunsaturated fat (6–11%) [c]	26 (43.3)	20 (47.6)	6 (33.3)	25 (44.6)	9 (42.9)	16 (45.7)	22 (41.5)	8 (47.1)	14 (38.9)
Saturated fat (< 10%) [c]	38 (63.3)	26 (61.9)	12 (66.7)	35 (62.5)	12 (57.1)	23 (65.7)	29 (54.7)	10 (58.8)	19 (52.8)
Cholesterol (<300 mg) [c] Mean SD	294.1 90.3	282.4 74.2	321.6 117.9	199.3 71.7	200.6 54.8	198.5 81.0	269.9 117.6	270.8 107.4	269.5 123.5

Treatment time points: T0, before or at beginning of treatment; T1, middle of treatment; T2, final of treatment; ONS, oral nutritional supplementation: without or with use; [a] AMDR, acceptable macronutrient distribution range [32]; [b] Reference for minimum recommended intake: 25 kcal/kg/day and 1 g protein/kg/day [33]; [c] [34]. Cholesterol intake should be minimized while consuming a nutritionally adequate diet.

Table 4. Prevalence of nutrient intake inadequacy using the estimated average requirement (EAR) method as the cut-off point, and comparison of intake with adequate intake (AI) in head and neck cancer patients during (chemo)radiotherapy.

Nutrients	Sex	Age Group (years)	DRI (EAR)	Prevalence of Inadequacy (%)								
					T0			T1			T2	
				Total (n = 60)	Without ONS (n = 42)	With ONS (n = 18)	Total (n = 56)	Without ONS (n = 21)	With ONS (n = 35)	Total (n = 53)	Without ONS (n = 17)	With ONS (n = 36)
Calcium (mg)	Male	31–70	800	62.17	79.67	40.13	30.50	51.20	21.48	28.43	75.17	14.23
		>70	1000	77.94	93.19	53.98	46.41	77.04	34.83	42.07	93.32	25.46
	Female	31–50	800	62.17	79.67	40.13	30.50	51.20	21.48	28.43	75.17	14.23
		>51	1000	77.94	93.19	53.98	46.41	77.04	34.83	42.07	93.32	25.46
Iron (mg)	Male	>31	6	21.77	23.27	17.36	17.11	18.14	14.01	15.62	27.09	9.51
	Female	31–50	8.1	36.69	42.07	27.76	25.46	33.72	20.61	23.89	47.61	15.62
		>51	5	16.11	16.35	13.35	13.79	12.71	11.51	12.51	18.94	7.35
Phosphorus (mg)	Male	>31	580	7.49	6.68	7.08	6.18	5.26	6.18	2.87	10.56	0.52
	Female	>31	580	7.49	6.68	7.08	6.18	5.26	6.18	2.87	10.56	0.52
Magnesium (mg)	Male	>31	350	86.86	95.35	70.88	69.50	94.63	59.48	79.10	97.98	69.50
	Female	>31	265	61.41	71.90	47.61	43.25	65.91	35.94	43.64	77.94	28.77
Niacin (mg)	Male	>31	12	26.43	30.15	18.41	24.20	33.36	18.94	8.85	24.20	2.33
	Female	>31	11	21.48	24.83	14.46	20.05	28.43	15.15	5.26	17.62	1.02
Riboflavin (mg)	Male	>31	1.1	45.62	60.26	24.20	24.20	35.20	17.36	22.96	51.60	12.92
	Female	>31	0.9	33.00	43.25	16.11	16.11	24.83	10.93	16.60	34.83	8.69
Thiamine (mg)	Male	>31	1.0	42.47	48.01	30.85	29.12	38.21	24.20	16.35	36.32	6.94
	Female	>31	0.9	34.46	38.97	24.83	23.89	30.15	19.77	11.90	28.77	4.36
Vitamin B6 (mg)	Male	31–50	1.1	84.61	96.86	63.31	49.60	81.33	39.36	49.20	87.29	35.94
		>51	1.4	96.08	99.83	80.78	66.28	95.73	54.78	73.57	98.50	62.55
	Female	31–50	1.1	84.61	96.86	63.31	49.60	81.33	39.36	49.20	87.29	35.94
		>51	1.3	93.45	99.49	75.80	61.03	92.51	49.60	65.91	96.64	53.59

Table 4. *Cont.*

Vitamin C (mg)	Male	>31	75	39.74	49.60	27.43	26.43	37.07	20.90	32.64	50.00	27.09
	Female	>31	60	31.56	36.32	22.36	19.22	21.48	15.15	25.14	35.94	20.61
Zinc (mg)	Male	>31	9.4	12.71	14.92	7.64	26.11	30.15	24.20	21.19	42.47	11.51
	Female	>31	6.8	2.81	3.75	1.13	12.51	13.79	11.70	11.51	26.43	5.05
			DRI (AI)	**Consumption Comparison with AI-Below Recommendation *n* (%)**								
Fiber (g)	Male	31–50 >51	38 30	9(15.0) 36 (60.0)	6(14.3) 24(57.1)	3(16.7) 12(66.7)	9(16.1) 35(62.5)	4(19.0) 12(57.1)	5(14.3) 23(65.7)	8(15.1) 32(60.4)	3(17.6) 9(52.9)	5(13.9) 23(63.9)
	Female	31–50 >51	25 21	3(5.0) 6(10.0)	3(7.1) 4(9.5)	0(0.0) 2(11.1)	2(3.6) 6(10.7)	2(9.5) 1(4.8)	0(0.0) 5(14.3)	3(5.7) 7(13.2)	2(11.8) 2(11.8)	1(2.8) 5(13.9)
Manganese (mg)	Male	>31	2.3	41(68.3) [a]	28(66.7)	13(72.2)	23(41.1) [a]	10(47.6)	13(37.1)	16(30.2) [a]	10(58.8)	6(16.7)
	Female	>31	1.8	6(10.0) [a]	4(9.5)	2(11.1)	6(10.7) [a]	4(19.0)	2(5.7)	6(11.3) [a]	3(17.6)	3(8.3)
Potassium (mg)	Male	>31	3400	21(35.0) [b]	13(31.0)	8(44.4)	20(35.7) [b]	7(33.3)	13(37.1)	11(20.8) [b]	8(47.1)	3(8.3)
	Female	>31	2600	3(5.0) [b]	3(7.1)	0(0.0)	4(7.1) [b]	2(9.5)	2(5.7)	1(1.9) [b]	1(5.9)	0(0.0)
Sodium (mg)	Male	>31	1500	0(0.0) [b]	0(0.0)	0(0.0)	4(7.1) [b]	2(9.5)	2(5.7)	2(3.8) [b]	1(5.9)	1(2.8)
	Female	>31	1500	0(0.0) [b]	0(0.0)	0(0.0)	1(1.8) [b]	0(0.0)	1(2.9)	0(0.0) [b]	0(0.0)	0(0.0)

Treatment time points: T0, before or at beginning of treatment; T1, middle of treatment; T2, final of treatment; DRI, Dietary Reference Intake [32]; EAR, estimated average requirement; AI, adequate intake, ONS, oral nutritional supplementation: without or with use. [a] Below UL (tolerable upper intake level) of 11 mg. [b] UL not determined due to lack of toxicological indicator specific to excessive potassium and sodium intake.

In Table 4, we observed that even though there was an increase in micronutrient intake during treatment with ONS use (Table 2), it was not enough to ensure adequacy compared to EAR. Inadequacy prevalence was lower in those who used ONS compared with those who did not. However, a high prevalence of inadequacy was observed, mainly for calcium, magnesium, and vitamin B6 intake, which was almost 100% for those who did not use ONS. Compared with AI, male patients presented a higher percentage of values below the recommended levels for fiber and manganese intake.

Regarding the initial nutritional status (six months before the treatment), the patients suffered negative mean BWC, and the overweight patients suffered a higher weight accumulated deficit with a delta of −15 kg (Figure 1A). Furthermore, those who used ONS showed less weight loss, except of the overweight patients (Figure 1B).

There was a decreasing trend of energy (kcal/kg/day, Figure 2A–C) and protein intake (g/kg/day, Figure 2D–F) comparing the underweight with overweight patients regardless of ONS use. Only the underweight group with and without ONS use showed an important difference considering the energy in T1 and protein consumption in T2 (Figure 2B,F).

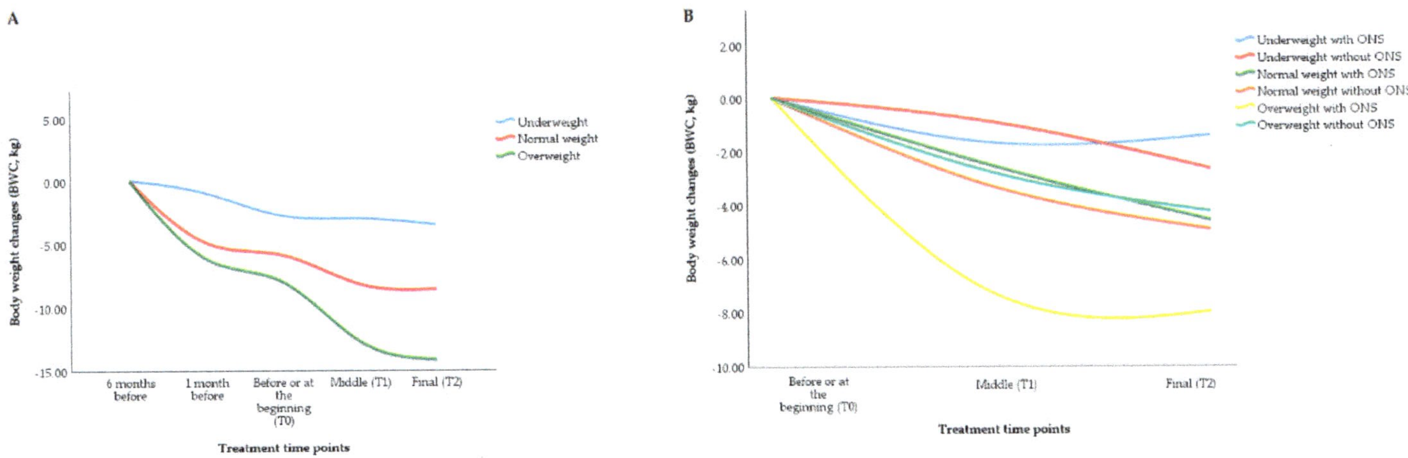

Figure 1. BWC by BMI groups. (**A**) BWC considering 5 time points (two points before treatment and the three treatment time points) for the patients categorized by habitual BMI (kg/m^2). (**B**) BWC for the patients considering three time points (the three treatment time points) categorized by T0 BMI (kg/m^2) and use or not of ONS. Abbreviations: BWC, body weight changes; BMI, body mass index; ONS, oral nutritional supplementation.

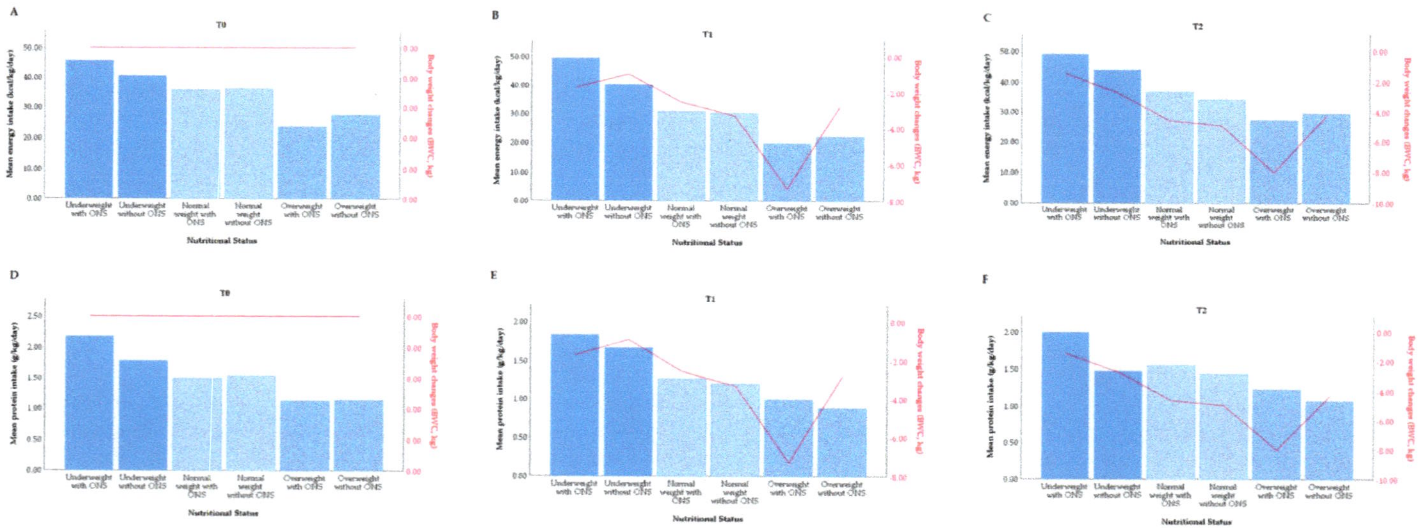

Figure 2. Mean energy and protein intake and BWC by nutritional status. (**A**) Mean energy intake and BWC considering three time points by nutritional status in T0. (**B**) Mean energy intake and BWC considering three time points by nutritional status in T1. (**C**) Mean energy intake and BWC considering three time points by nutritional status in T2. (**D**) Mean protein intake and BWC considering three time points by nutritional status in T0. (**E**) Mean protein intake and BWC considering three time points by nutritional status in T1. (**F**) Mean protein intake and BWC considering three time points by nutritional status in T2. Abbreviations: BWC, body weight changes; ONS, oral nutritional supplementation.

4. Discussion

A significant reduction in energy, macronutrients, and cholesterol intake at the middle of treatment (first month) and the return of these consumption levels at the end of treatment (second month) were observed. Regarding micronutrients, the majority of patients increased their intake from the beginning to the middle of treatment and the values were maintained until the end due to ONS use. However, energy, macro-, and micronutrient inadequacy prevalence was high in all time points, especially in patients who did not use ONS. Finally, overweight patients suffered a higher weight accumulated deficit compared to other BMI categories. In general, those who used ONS showed less weight loss, except for overweight patients, and only the underweight group with and without ONS use showed an important difference considering the energy in T1 and protein consumption in T2.

Recent studies in the literature have evaluated dietary intake after treatment (long-term), but not during the treatment. They assessed at diagnosis and post induction chemotherapy, after RT, from one and three months after the end of treatment [22], or at baseline and post-treatment (after 4–6 weeks of RT and/or CT, and follow-up (8–10 weeks after completion of treatment) [10]. The difference in our study is that we identified the treatment time points at which there was the greatest impact on dietary intake and showed the high prevalence of the inadequacy of energy and nutrients (short-term). This is important because the greater impact during treatment can cause negative consequences that can be predicted if the patient is monitored, even if the decrease in food intake and weight values is expected. In addition, we used GEE modeling to estimate not only the effects of treatment time points on dietary intake, but also the contribution of ONS on nutrient intake. Besides the lack of studies on dietary intake and HNC in the literature, there is also no data on the prevalence of inadequacy. These data are important to assess whether the amounts of micronutrients ingested are sufficient. This can contribute to an adequate nutritional approach and may reduce the risk of adverse health outcomes in these patients, since, as evidenced in this study, even having increased the intake of micronutrients with ONS use, patients consumed micronutrients in quantities below the recommendation. Furthermore, to the best of our knowledge, no study has adjusted for intra-individual variability of values related to energy and nutrient consumption in order to better estimate individual intake, as suggested by Nusser et al. [29]; nor have they adjusted for the total energy of the sample by the residual method due to the association between energy and most nutrients, as recommended in the literature [30]. Thus, the studies that evaluated the food intake of patients with HNC have not performed these necessary analyses.

As symptoms presented by HNC patients such as xerostomia, dysgeusia, dysphagia, mucositis, and thick saliva [36,37] can limit oral intake [22] and lead to changes in food consistency, our results for macronutrients were expected. A lower macronutrient intake is expected when patients opt for pasty, liquid, or mild food [9], impacting in particular fiber content. Additionally, an increase in intake of soup and foods prepared with milk has been reported [10,36]. The same decreased intake seem in studies that assessed the dietary intake in the long-term (at diagnosis, post treatment, and follow-up) have found a reduction in energy and protein intake in HNC patients [7,22]. Adequate protein intake can minimize the severity of oral mucositis in patients with HNC undergoing RT due to the ability of protein to maintain integrity or repair mucosal lesions [38]. On the other hand, a low protein intake can increase the risk of fatigue and mortality in advanced cancer patients undergoing CT [39].

Cancer patients present impaired macronutrient metabolism due to systemic inflammation, which can lead to altered protein turnover, loss of fat and muscle mass, increased production of acute phase proteins, insulin resistance, glucose intolerance, and increase or maintenance of lipid oxidation capacity [33]. Therefore, impairment of these metabolic pathways has a negative effect on clinical outcome and macronutrients are needed for bodily maintenance and better response to treatment. However, in this study, although patients who used ONS consumed more protein than those who did not, protein intake in grams presented a higher percentage of inadequacy at T1. Therefore, nutritional intervention with complete assessment is extremely important to make changes in the diet of these patients, avoiding recommendations directed only at increasing energy and protein intake without assessing food quality. Furthermore, it is also important to assess the amount ingested and not only the use or not of ONS, or indicate its use without providing guidance to patients, because, depending on the severity and if the ONS does not have the amount of proteins necessary to minimize the impacts, protein modules can be used in order to achieve nutritional requirements as soon as possible.

Unlike macronutrients, the mean intake of micronutrients increased over the duration of treatment. This suggests that ONS may have been determinant in increasing micronutrient intake, although not at a high enough level to ensure adequacy. A high prevalence of energy and nutrient inadequacy was observed despite ONS use, mainly for calcium, magnesium, and vitamin B6. This reveals that HNC patients require nutritional intervention with a special attention to the quantification of food intake in order to estimate possible deficits and achieve adequate levels of macro- and micronutrients through prophylactic ONS prescription. Previous studies have highlighted the importance of vitamin B6 as a protective factor against the development of cancer [40], antioxidant effects [41], and increased immune response [42].

In addition, magnesium participates in energy metabolism, protein synthesis, and plays an important physiological role in organs such as the brain and heart [43]. In relation to low micronutrient intake, there is a high prevalence of vitamin D deficiency among HNC patients [44]. This deficiency has been linked to an increased risk of postoperative hypocalcemia in patients undergoing total thyroidectomy [45]. Low levels of calcium, vitamin E, and folate intake were also found in patients with HNC [10,44]. So, while adequate levels of some micronutrients can be obtained from a healthy diet, inadequate dietary intake of others can lead to negative health consequences.

Moreover, our study also showed an increase in ONS usage by time points. In other words, ONS complements other food intake, being an alternative route to achieving the recommended levels of micronutrient intake [10]. Recent studies have shown less CRT-related toxicity, better weight maintenance, and tolerance to treatment with nutritional counseling using ONS [15,19,46].

Although micronutrient intake and ONS use increased during treatment, the number of malnourished patients also increased and overweight patients decreased between time points. Thus, a decrease in BMI and an increase in malnutrition led to more ONS prescription and to more frequent use, since these patients do not use ONS prophylactically. This situation may be due to late diagnosis or advanced age as well as low dietary intake and treatment side effects.

According to the ESPEN guidelines, nutritional intervention including ONS provision is recommended to ensure adequate dietary intake, prevent weight loss, and avoid treatment discontinuation in HNC patients [33]. However, as previously mentioned, the majority of patients did not receive nutritional counseling at the beginning of treatment and approximately one third did not receive it at any time point. Thus, nutritional interventions through individualized dietary counseling can positively influence long-term outcomes related to quality of life and nutritional status [14,21,47].

Similar to other studies [15,16,19], we observed that those who used ONS showed less weight loss (except of the overweight) and consumed more protein (0.14 g/kg/day). Additionally, malnourished patients lost less weight, while overweight patients lost more weight, showing that the initial BMI defines the BWC. An additional finding was an accumulated deficit of weight loss that was found before the start of treatment. Corroborating these findings, Orell et al. [16] found that overweight patients lost more weight due to symptoms such as anorexia and nausea. According to the Aspen Guideline [48], critically ill patients with obesity have greater complications when compared to patients with normal weight classified by BMI and have an indication for assessment and early nutritional support. These results are important because they show that there is a significant cumulative weight loss before treatment and that overweight patients may lose more weight not only due to the presence of the disease and the symptoms presented, but also because they are not receiving nutritional counseling according to their needs. Therefore, more attention should be given to overweight patients, since they are often neglected and have been little explored by studies, despite presenting more mass and worse deficit.

Thus, this study showed how much macronutrient intake decreased during treatment and that, even consuming ONS, micronutrient intake is below that recommended. Furthermore, it also demonstrated that the patients suffered negative mean BWC and the overweight patients had the highest accumulated weight deficit. This reinforces the importance of following the Guidelines for HNC patients [13] in which they guide the weekly consultation with a dietitian during treatment to obtain better results. Therefore, we strongly recommend that dietary counseling consultations are routinely provided following cancer diagnosis in order to carry out more specific nutritional orientations such as food recipes fortified with ONS, which will improve caloric intake and maintain adequate intake, avoiding a deficit in nutrient consumption that will cause weight loss and increases in malnutrition, thereby helping the recovery of HNC patients.

This study has some limitations. 24HR was used, and although it is the most accurate tool for dietary intake analysis, it may present a memory bias because it depends on an individual's ability to accurately recall their food intake. However, in order to minimize this limitation, interviews were conducted by trained dietitians and there was standardization at the time point of collection of 24HR as well as the use of the multiple pass method and typing in order to obtain more reliable results. Moreover, nutrient intake was adjusted for intra-individual variability and for energy intake in order to present intake estimates as precisely as possible. Finally, we did not separate ONS from enteral nutrition because there was no difference in consumption. Therefore, what was consumed in both systems was evaluated.

The study also has strengths. It evaluates the short-term impact on dietary intake, which is during treatment, and assesses the prevalence of inadequate micronutrient intake. This is clinically important since we can minimize negative health outcomes such as malnutrition, delay in post-treatment recovery, longer convalescence, and other long-term impacts such as decreased quality of life and mortality.

5. Conclusions

Head and neck cancer patients showed energy and macronutrient intake decreased at the middle of treatment and the increased micronutrient intake due to ONS use. Despite this, the prevalence of inadequate energy and nutrient intake, particularly for calcium, magnesium, and vitamin B6 was high in all time points even with ONS use, but proved worse for those who did not use ONS. Furthermore, overweight patients suffered a higher weight accumulated deficit compared to other BMI categories. Patients on ONS showed a lower weight deficit. Therefore, we strongly recommend initiating nutritional counseling from diagnosis to optimize macronutrient intake in conjunction with prophylactic ONS prescription to adjust micronutrient intake and minimize the weight loss, making it possible to prevent worse prognosis and nutritional status.

Author Contributions: Conceptualization, I.B.F., E.d.N.S.L., Y.C.d.P.M., and G.d.G.P.; Formal analysis, I.B.F., E.N.S.L., C.A.G., Y.C.d.P.M., and G.d.G.P.; Investigation, I.B.F and E.N.S.L.; Writing—original draft preparation, I.B.F., E.N.S.L., P.P.L.C., C.A.G., Y.C.d.P.M., and G.d.G.P.; Writing—review and editing, I.B.F., E.N.S.L., P.P.L.C., C.A.G., Y.C.d.P.M., and G.d.G.P.; Visualization, I.B.F. All authors have read and agreed to the published version of the manuscript.

Funding: This research received no external funding.

Acknowledgments: We are grateful to all volunteers who participated in this study. We also would like to thank the Coordination of Improvement of Higher Education Personnel (Coordenação de Aperfeiçoamento de Pessoal de Nível Superior (CAPES), Brazil for their support with the publishing fee.

Conflicts of Interest: The authors declare no conflict of interest.

References

1. Wild, C. *World Health Organization World Cancer Report 2014 (ePUB)*; World Health Organization: Geneva, Switzerland, 2014; Volume 7, pp. 418–419. [CrossRef]
2. Dechaphunkul, T.; Martin, L.; Alberda, C.; Olson, K.; Baracos, V.; Gramlich, L. Malnutrition assessment in patients with cancers of the head and neck: A call to action and consensus. *Crit. Rev. Oncol. Hematol.* **2013**, *88*, 459–476. [CrossRef] [PubMed]
3. Bressan, V.; Stevanin, S.; Bianchi, M.; Aleo, G.; Bagnasco, A.; Sasso, L. The effects of swallowing disorders, dysgeusia, oral mucositis and xerostomia on nutritional status, oral intake and weight loss in head and neck cancer patients: A systematic review. *Cancer Treat. Rev.* **2016**, *45*, 105–119. [CrossRef] [PubMed]
4. García-Peris, P.; Parón, L.; Velasco, C.; de la Cuerda, C.; Camblor, M.; Bretón, I.; Herencia, H.; Verdaguer, J.; Navarro, C.; Clave, P. Long-term prevalence of oropharyngeal dysphagia in head and neck cancer patients: Impact on quality of life. *Clin. Nutr.* **2007**, *26*, 710–717. [CrossRef] [PubMed]
5. Kenway, N.; Leung, S.F.; Johnson, P.J.; Woo, J. Nutritional Consequences of Radiotherapy in Nasopharynx Cancer Patients. *Nutr. Cancer* **2004**, 37–41. [CrossRef]
6. Langius, J.A.E.; Kruizenga, H.M.; Uitdehaag, B.M.J.; Langendijk, J.A.; Doornaert, P.; Leemans, C.R.; Weijs, P.J.M. Resting energy expenditure in head and neck cancer patients before and during radiotherapy. *Clin. Nutr.* **2012**, *31*, 549–554. [CrossRef]
7. van den Berg, M.G.A.; Rasmussen-Conrad, E.L.; Gwasara, G.M.; Krabbe, P.F.M.; Naber, A.H.J.; Merkx, M.A. A prospective study on weight loss and energy intake in patients with head and neck cancer, during diagnosis, treatment and revalidation. *Clin. Nutr.* **2006**, *25*, 765–772. [CrossRef]
8. Marshall, K.M.; Loeliger, J.; Nolte, L.; Kelaart, A.; Kiss, N.K. Prevalence of malnutrition and impact on clinical outcomes in cancer services: A comparison of two time points. *Clin. Nutr.* **2018**, 1–8. [CrossRef]
9. Chavoni, R.; Silva, P.; Ramos, G. Diagnóstico nutricional de pacientes do serviço de cabeça e pescoço e sua relação com a disfagia em um hospital oncológico do Paraná. *Rev. Bras. Cir. Cabeça Pescoço* **2014**, *43*, 35–41.
10. Nejatinamini, S.; Kubrak, C.; Álvarez-Camacho, M.; Baracos, V.E.; Ghosh, S.; Wismer, W.V.; Mazurak, V.C. Head and Neck Cancer Patients Do Not Meet Recommended Intakes of Micronutrients without Consuming Fortified Products. *Nutr. Cancer* **2018**, *70*, 474–482. [CrossRef]
11. Medeiras, F.P.P.; Martinez, C.E.; Cardoso, S.D.S. Estado Nutricional E Ingestão Alimentar De Pacientes Com Câncer De Cabeça E Pescoço Submetidos a Tratamento Oncológico. *Arq. Ciências Saúde* **2016**, *23*, 43. [CrossRef]
12. Kennedy, D.O. B vitamins and the brain: Mechanisms, dose and efficacy—A review. *Nutrients* **2016**, *8*, 68. [CrossRef] [PubMed]
13. Talwar, B.; Donnelly, R.; Skelly, R.; Donaldson, M. Nutritional management in head and neck cancer: United Kingdom National Multidisciplinary Guidelines. *J. Laryngol. Otol.* **2016**, *130*, S32–S40. [CrossRef] [PubMed]
14. Ravasco, P.; Monteiro-Grillo, I.; Vidal, P.M.; Camilo, M.E. Impact of nutrition on outcome: A prospective randomized controlled trial in patients with head and neck cancer undergoing radiotherapy. *Head Neck* **2005**, *27*, 659–668. [CrossRef] [PubMed]
15. Yang, Y.C.; Lee, M.S.; Cheng, H.L.; Chou, H.Y.; Chan, L.C. More Frequent Nutrition Counseling Limits Weight Loss and Improves Energy Intake During Oncology Management: A Longitudinal Inpatient Study in Taiwan. *Nutr. Cancer* **2019**, *71*, 452–460. [CrossRef]
16. Orell, H.; Schwab, U.; Saarilahti, K.; Österlund, P.; Ravasco, P.; Mäkitie, A. Nutritional counseling for head and neck cancer patients undergoing (chemo) radiotherapy—A prospective randomized trial. *Front. Nutr.* **2019**, *6*. [CrossRef]
17. Kang, W.-X.; Li, W.; Huang, S.-G.; Dang, Y.; Gao, H. Effects of nutritional intervention in head and neck cancer patients undergoing radiotherapy: A prospective randomized clinical trial. *Mol. Clin. Oncol.* **2016**, *5*, 279–282. [CrossRef]
18. Van Den Berg, M.G.A.; Rasmussen-Conrad, E.L.; Wei, K.H.; Lintz-Luidens, H.; Kaanders, J.H.A.M.; Merkx, M.A.W. Comparison of the effect of individual dietary counselling and of standard nutritional care on weight loss in patients with head and neck cancer undergoing radiotherapy. *Br. J. Nutr.* **2010**, *104*, 872–877. [CrossRef]

19. Cereda, E.; Cappello, S.; Colombo, S.; Klersy, C.; Imarisio, I.; Turri, A.; Caraccia, M.; Borioli, V.; Monaco, T.; Benazzo, M.; et al. Nutritional counseling with or without systematic use of oral nutritional supplements in head and neck cancer patients undergoing radiotherapy. *Radiother. Oncol.* **2017**, *126*, 81–88. [CrossRef]
20. Brown, T.; Ross, L.; Jones, L.; Hughes, B.; Banks, M. Nutrition outcomes following implementation of validated swallowing and nutrition guidelines for patients with head and neck cancer. *Support. Care Cancer* **2014**, *22*, 2381–2391. [CrossRef]
21. Langius, J.A.E.; Zandbergen, M.C.; Eerenstein, S.E.J.; van Tulder, M.W.; Leemans, C.R.; Kramer, M.H.H.; Weijs, P.J.M. Effect of nutritional interventions on nutritional status, quality of life and mortality in patients with head and neck cancer receiving (chemo)radiotherapy: A systematic review. *Clin. Nutr.* **2013**, *32*, 671–678. [CrossRef]
22. Arribas, L.; Hurtós, L.; Taberna, M.; Peiró, I.; Vilajosana, E.; Lozano, A.; Vazquez, S.; Mesia, R.; Virgili, N. Nutritional changes in patients with locally advanced head and neck cancer during treatment. *Oral Oncol.* **2017**, *71*, 67–74. [CrossRef] [PubMed]
23. WMA World Medical Association. Declaration of Helsinki: Ethical Principles for Medical Research Involving Human Subjects. *Bull. World Health Organ.* **2001**, *79*, 373.
24. Chafidz, A.; Kaavessina, M.; Al-Zahrani, S.; Al-Otaibi, M.N. AJCC Cancer Stating Manual, 7th edition. American Joint Committee on Cancer. Available online: https://cancerstaging.org/references-tools/deskreferences/pages/default.aspx (accessed on 1 January 2018).
25. World Health Organization. Physical Status: The use and interpretation of anthropometry. In *Report of a WHO Expert Committee*; WHO Technical Report Series 854; WHO: Geneva, Switzerland, 1995; Available online: https://apps.who.int/iris/handle/10665/37003 (accessed on 2 January 2016).
26. Moshfegh, A.J.; Rhodes, D.G.; Baer, D.J.; Murayi, T.; Clemens, J.C.; Rumpler, W.V.; Paul, D.R.; Sebastian, R.S.; Kuczynski, K.J.; Ingwersen, L.A.; et al. The US Department of Agriculture Automated Multiple-Pass Method reduces bias in the collection of energy intakes. *Am. J. Clin. Nutr.* **2008**, *88*, 324–332. [CrossRef]
27. Universidade Estadual de Campinas. *TACO Tabela Brasileira de Composição de Alimentos*; Universidade Estadual de Campinas: Campinas, Brazil, 2011. [CrossRef]
28. USDA United States Dietetic Association. *Dietary Guidelines for Americans*; US Department of Health and Human Services, US Department of Agriculture: Davis, CA, USA, 2010. [CrossRef]
29. Nusser, S.; Carriquiry, A.; Dodd, K.; Fuller, W. A semiparametric transformation approach to estimating usual daily intake distributions. *J. Am. Stat. Assoc.* **1996**, *91*, 1440–1449. [CrossRef]
30. Willett, W.; Howe, G.; Kushi, L. Adjustment for total energy intake in epidemiologic studies. *Am. J. Clin. Nutr.* **1997**, *65*, 1220S–1228S. [CrossRef] [PubMed]
31. Beaton, G.H. Approaches to analysis of dietary data: Relationship between planned analyses and choice of metodology. *Am. J. Clin. Nutr.* **1994**, *59*, 253S–261S. [CrossRef]
32. IOM Institute of Medicine. *Dietary Reference Intakes: Applications in Dietary Planning*; 2003; The National Academies Press: Washington, DC, USA.
33. Arends, J.; Bachmann, P.; Baracos, V.; Barthelemy, N.; Bertz, H.; Bozzetti, F.; Fearon, K.; Hütterer, E.; Isenring, E.; Kaasa, S.; et al. ESPEN guidelines on nutrition in cancer patients. *Clin. Nutr.* **2017**, *36*, 11–48. [CrossRef]
34. FAO, Food and Agriculture Organization of the United Nations. *Fats and Fatty Acids in Human Nutrition: Report of an Expert Consultation*; FAO Food and Nutrition Paper 91; FAO: Rome, Italy, 2010; Volume 91, ISBN 9789251067338.
35. Wang, Y.G.; Fu, L. Selection of working correlation structure in generalized estimating equations. *Stat. Med.* **2017**, *36*, 2206–2219. [CrossRef]
36. Álvarez-Camacho, M.; Martínez-Michel, L.; Gonella, S.; Scrimger, R.A.; Chu, K.P.; Wismer, W.V. Physical symptom burden of post-treatment head and neck cancer patients influences their characterization of food: Findings of a repertory grid study. *Eur. J. Oncol. Nurs.* **2016**, *22*, 54–62. [CrossRef]
37. Moroney, L.B.; Helios, J.; Ward, E.C.; Crombie, J.; Pelecanos, A.; Burns, C.L.; Spurgin, A.L.; Blake, C.; Kenny, L.; Chua, B.; et al. Helical intensity-modulated radiotherapy with concurrent chemotherapy for oropharyngeal squamous cell carcinoma: A prospective investigation of acute swallowing and toxicity patterns. *Head Neck* **2018**, *40*, 1955–1966. [CrossRef]

38. Zahn, K.; Wong, G.; Bedrick, E.; Poston, D.; Schroeder, T.; JE, B. Relationship of protein and calorie intake to the severity of oral mucositis in patients with head and neck cancer receiving radiation therapy. *Head Neck* **2012**, *34*, 655–662. [CrossRef] [PubMed]
39. Stobäus, N.; Müller, M.J.; Küpferling, S.; Schulzke, J.D.; Norman, K. Low Recent Protein Intake Predicts Cancer-Related Fatigue and Increased Mortality in Patients with Advanced Tumor Disease Undergoing Chemotherapy. *Nutr. Cancer* **2015**, *67*, 818–824. [CrossRef] [PubMed]
40. Larsson, S.C.; Orsini, N.; Wolk, A. Vitamin B6 and Risk of Colorectal Cancer: A Meta-Analysis of Prospective Studies. *JAMA* **2010**, *303*, 1077. [CrossRef] [PubMed]
41. Merigliano, C.; Mascolo, E.; Burla, R.; Saggio, I.; Vernì, F. The Relationship between Vitamin B6, Diabetes and Cancer. *Front. Genet.* **2018**, *9*, 1–5. [CrossRef]
42. Sujol, G.; Docquier, A.; Boulahtouf, A.; Castet-Nicolas, A.; Cavaillès, V. Vitamine B6 et cancer: Des données cliniques aux mécanismes moléculaires. *Bull. Cancer* **2011**, *98*, 1201–1208. [CrossRef]
43. de Baaij, J.H.F.; Hoenderop, J.G.J.; Bindels, R.J.M. Magnesium in Man: Implications for Health and Disease. *Physiol. Rev.* **2014**, *95*, 1–46. [CrossRef]
44. Nejatinamini, S.; Debenham, B.J.; Clugston, R.D.; Mawani, A.; Parliament, M.; Wismer, W.V.; Mazurak, V.C. Poor vitamin status is associated with skeletal muscle loss and mucositis in head and neck cancer patients. *Nutrients* **2018**, *10*, 1236. [CrossRef]
45. Alkhalili, E.; Ehrhart, M.; Ayoubieh, H.; Burge, M. Does pre-operative vitamin d deficiency predict postoperative hypocalcemia after thyroidectomy? *Endocr. Pract.* **2017**, *23*, 5–9. [CrossRef]
46. Valentini, V.; Marazzi, F.; Bossola, M.; Micciché, F.; Nardone, L.; Balducci, M.; Dinapoli, N.; Bonomo, P.; Autorino, R.; Silipigni, S.; et al. Nutritional counselling and oral nutritional supplements in head and neck cancer patients undergoing chemoradiotherapy. *J. Hum. Nutr. Diet.* **2012**, *25*, 201–208. [CrossRef]
47. Ravasco, P.; Monteiro Grillo, I.; Camilo, M. Cancer wasting and quality of life react to early individualized nutritional counselling! *Clin. Nutr.* **2007**, *26*, 7–15. [CrossRef]
48. Choban, P.; Dickerson, R.; Malone, A.; Worthington, P.; Compher, C.A. Clinical guidelines: Nutrition support of hospitalized adult patients with obesity. *J. Parenter. Enter. Nutr.* **2013**, *37*, 714–744. [CrossRef] [PubMed]

nutrients

MDPI

Article

Comparison between Percutaneous Gastrostomy and Self-Expandable Metal Stent Insertion for the Treatment of Malignant Esophageal Obstruction, after Propensity Score Matching

Joo Hye Song [1], Jaehyun Ko [1], Yang Won Min [1,*], Kyunga Kim [2], Hyuk Lee [1], Byung-Hoon Min [1], Jun Haeng Lee [1], Poong-Lyul Rhee [1] and Jae J. Kim [1]

1 Department of Medicine, Samsung Medical Center, Sungkyunkwan University School of Medicine, Seoul 06351, Korea; joohye.song@samsung.com (J.H.S.); jaehyun0820.ko@samsung.com (J.K.); lhyuk.lee@samsung.com (H.L.); jason.min@samsung.com (B.-H.M.); jh2145.lee@samsung.com (J.H.L.); pl.rhee@samsung.com (P.-L.R.); jaej.kim@samsung.com (J.J.K.)

2 Statistics and Data Center, Research Institute for Future Medicine, Samsung Medical Center, Seoul 06351, Korea; kyunga.j.kim@samsung.com

* Correspondence: yangwonee@gmail.com; Tel.: +82-2-3410-3409

Received: 9 August 2020; Accepted: 7 September 2020; Published: 10 September 2020

Abstract: Background: The outcomes of the two procedures; self-expandable metal stent (SEMS) insertion and percutaneous gastrostomy (PG) feeding procedures, used in patients with malignant esophageal obstruction, are still controversial. We aimed to compare the outcomes between the two procedures, following propensity score (PS) matching. Methods: We retrospectively reviewed 568 esophageal cancer patients who underwent SEMS insertion (stent group) or PG (gastrostomy group) at the Samsung Medical Center between January 1996 and December 2018. Procedures for reasons other than malignant obstruction were excluded. We analyzed the datasets after PS matching. Primary outcomes were the post-procedural nutritional status, and need for additional intervention (AI). The secondary outcome was overall survival (OS). Results: In a matched cohort, the gastrostomy group showed less decrease in albumin level after the procedure (-0.15 ± 0.57 vs. stent group; 0.41 ± 0.59, $p = 0.021$). The gastrostomy group required less need for, and number of, AIs (2.1% vs. stent group; 23.4%, $p < 0.001$ and 0.04 ± 0.25 vs. stent group; 0.31 ± 0.61, $p < 0.001$). After matching, there was no significant difference between the two groups in OS. However, PG was associated with OS based on multivariable analysis of the matched cohort (vs. stent group, hazard ratio 0.69, 95% confidence interval 0.5–0.95). Conclusions: PG tends to provide better post-procedure nutritional status than SEMS insertion in patients with malignant esophageal obstruction.

Keywords: esophageal neoplasm; self-expandable metallic stents (SEMS); gastrostomy; enteral nutrition; survival

1. Introduction

Esophageal cancer is the eighth most common cancer and sixth leading cause of cancer-related mortality worldwide [1]. More than 50% of cases of esophageal cancer are usually diagnosed at an advanced stage, and dysphagia is the most common symptom, which contributes to weight loss and malnutrition [2]. Self-expandable metal stent (SEMS) insertion was established as the standard treatment for patients with malignant esophageal obstruction [3]. However, SEMS insertion led to adverse outcomes, such as chest pain, fistula, and stent migration [4,5]. Percutaneous gastrostomy (PG) feeding was established as an alternative treatment for malignant esophageal obstruction, and several studies suggested that PG could provide stable nutritional status and a better quality of life (QoL) compared to stent insertion [6–11].

From a physiological point of view, stent insertion is an ideal procedure [4,12]. However, there were inconsistent outcomes and a lack of strong evidence supporting its implementation for optimal treatment of patients with malignant esophageal obstruction. Usually, multidisciplinary evaluation before deciding on the route of feeding is required [13]. When the management of malignant esophageal obstruction is planned by a physician, it is important to predict the nutritional outcomes after the procedure is conducted. This is mainly because esophageal cancer exhibits a high risk of malnutrition related to cancer cachexia and dysphagia [2].

Therefore, in our study, we aimed to compare the outcomes related to nutritional and survival benefits between SEMS insertion and PG feeding for patients with malignant esophageal obstruction by applying the propensity score matching method.

2. Materials and Methods

2.1. Study Population

We retrospectively reviewed 568 esophageal cancer patients who underwent fully covered SEMS insertion (stent) or PG (gastrostomy) at the Samsung Medical Center between January 1996 and December 2018. Exclusion criteria were as follows: (1) underwent procedures for reasons other than malignant obstruction (n = 106), such as stricture (n = 50) (due to radiation therapy (n = 12), endoscopic submucosal dissection (n = 8), or esophagectomy (n = 30)), fistula (n = 52), esophageal perforation (n = 1), and aspiration pneumonia (n = 3); (2) coexisting other malignancies (n = 65); and (3) others (n = 14), such as underwent the procedure previously at another hospital (n = 7), recurrent cancer after radiotherapy and esophagectomy (n = 2), lye stricture (n = 1), underwent jejunostomy prior to the procedure (n = 1), did not follow-up (n = 1), and tissue type of esophageal cancer was unclear or unclassified (n = 2). Finally, a total number of 383 patients, including 195 patients who underwent SEMS insertion and 188 patients who received PG feeding, were considered for further analysis in this study (Figure 1).

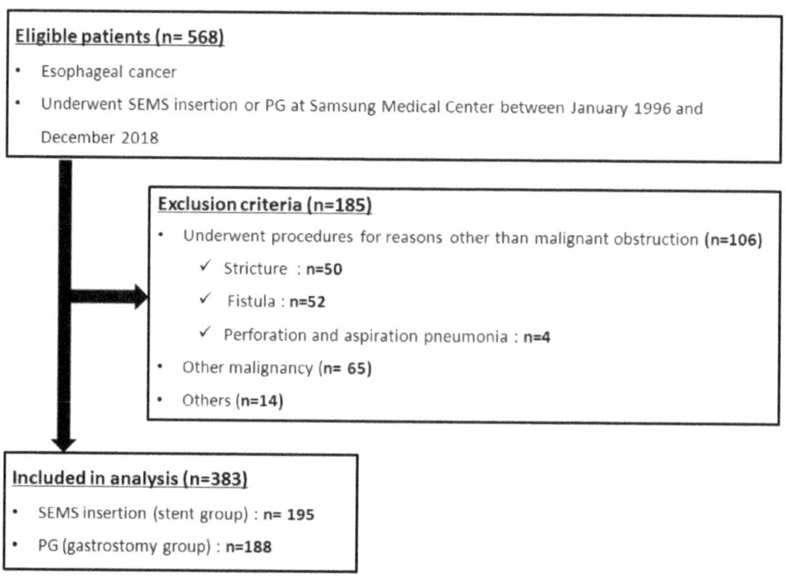

Figure 1. Flow chart of the study population. SEMS, self-expandable metal stent; PG, percutaneous gastrostomy.

On 15 May 2020, the Institutional Review Board of the Samsung Medical Center provided their approval to conduct this study (2020-05-018-001). The study protocol conformed to the ethical guidelines of the 1975 Declaration of Helsinki, as reflected in a prior approval by the institution's human research committee. The requirement for informed consent from patients was waived because only de-identified data routinely collected during hospital visits were used.

2.2. Indication of Procedure

At our hospital, difficulty in providing sufficient nutrition due to dysphagia was considered as an indication for performing either SEMS insertion or PG. When the attending physician chose the procedure type (stent or gastrostomy), the physician discussed with patients the known benefits and complications of each procedure based on general clinical experience. And finally, the physician decided the route for feeding. In case of cervical esophageal obstruction, the physician preferred PG to stent insertion due to the difficulty in maintaining the position of the stent. Conversely, when it was difficult to perform PG due to anatomical reasons, the physician preferred stent to PG. Endoscopic stent insertion was performed under conscious sedation using endoscopy. Once the obstruction was adequately delineated and guide wire access through the entire length of the obstruction obtained, the SEMS was advanced across the obstruction and deployed. The stent was deployed under a combination of endoscopic and fluoroscopic guidance. Oral feeding was possible from the day after stent insertion. PG was also performed under endoscopic (percutaneous endoscopic gastrostomy, PEG) or fluoroscopic guidance (percutaneous radiologic gastrostomy, PRG). Prophylactic intravenous antibiotics were administrated before the procedure. The PEG tube was inserted by the pull method. The puncture site was marked with endoscopic monitoring of the anterior gastric wall in the lower body by trans-illumination, and, following adequate local anesthesia, an appropriate initial incision was made and the puncture cannula was inserted under endoscopic control through the stomach. The PRG tube was inserted directly through the abdominal wall into the stomach by the push method. Enteral feeding was possible 24 h after the procedure, only if the patient had no abdominal pain and displayed normal bowel sound. About 50 mL of normal saline/hour was administrated three times, and then 50 mL of semi-fluid diet feeding (SFD) was administered. The SFD feeding reached 250 mL by gradual increases in the amount of feeding, and then soft blend diet feeding was initiated.

2.3. Outcome Measurement

Baseline characteristics were assessed retrospectively by reviewing electronic medical records as follows: age at diagnosis, gender, tumor stage according to the 7th edition of the American Joint Committee on Cancer, tumor histology, tumor location, length of obstruction by tumor (assessed by esophagography, endoscopy or computed tomography (CT)), history of chemotherapy, radiotherapy, and esophagectomy before the procedure [14]. We also assessed the procedure-related adverse events (such as tumor bleeding, fistula, perforation, and chest pain in the stent group, and PG site infection, peritonitis, and leakage in the gastrostomy group), the presence and number of additional interventions (AIs) (including stent insertion/repositioning, gastrostomy or removal of stent/gastrostomy due to adverse events), and the occurrence of aspiration pneumonia. We assessed the presence of chest pain by reviewing medical records where the patient had subjectively complained of chest pain, or the physician had carried out a procedure for chest pain. Aspiration pneumonia was defined as the combination of a history of aspiration according to the medical records and gravity-dependent opacity in a chest CT scan after the procedure. Furthermore, in order to evaluate the post-procedural nutritional status, body weight and the serum albumin level at baseline and 1 month after the procedure were assessed.

Primary outcomes were post-procedural nutritional status (a change in serum albumin level and body weight between baseline and 1 month after procedure), the need for and number of AIs, and the occurrence of procedure-related adverse events and aspiration pneumonia. The secondary outcome was overall survival (OS).

2.4. Statistical Analysis

Continuous variables are expressed as mean ± standard deviation, while categorical variables are presented as absolute values and percentages. Differences between continuous variables were analyzed using the unpaired Student's *t*-test and Wilcoxon rank sum test. The differences between categorical variables were analyzed using the $\chi2$ test and Fisher's exact test, accordingly. The propensity scores were estimated for age, gender, stage, length of obstruction and treatment before procedure using the parsimonious logistic regression model. The 1:1 matching without replacement was performed within 25% of standard deviation of log-transformed propensity scores, therefore the matched data were analyzed with exactly the same methods that can be used for the original data [15,16]. In the propensity score-matched cohort, the two groups were compared for the baseline characteristics, and the absolute standardized mean differences of variables were <0.2 to be balanced between two groups. OS was calculated using the Kaplan–Meier method and compared using the log-rank test. Cox hazard proportional models were used to examine the association of baseline characteristics with overall survival in the propensity score-matched cohort. Variables with a p-value < 0.2 in univariable analysis were later subjected to multivariable analysis. Differences with a p-value < 0.05 were considered statistically significant. All statistical analyses were performed using SPSS software version 25.0 for Windows (SPSS Inc., Chicago, IL, USA) and open source statistical language and platform, R, version 3.6.1 (R Development Core Team, Vienna, Austria), using the package "Matching."

3. Results

3.1. Baseline Characteristics of the Study Population

Baseline characteristics of patients with malignant esophageal obstruction who underwent stent insertion and gastrostomy are shown in Table 1. Before matching, the patients in the stent group (n = 195) were identified to be in a more advanced stage, with a lower number of cervical cancer cases, and received chemotherapy, radiotherapy, and esophagectomy less often than the patients in the gastrostomy group (n = 188) ($p < 0.001$ for all the characteristics). After propensity score matching, there was no significant difference between the stent and gastrostomy groups (Table 2).

Table 1. Comparison of baseline characteristics between patients with malignant esophageal obstruction undergoing gastrostomy and stent insertion, before propensity score matching (n = 383).

		Gastrostomy Group (n = 188)	Stent Group (n = 195)	p-Value
Age	year	63.80 ± 9.57	64.71 ± 10.36	0.37
Gender	Male	161 (85.6)	180 (92.3)	0.054
	Female	27 (14.4)	15 (7.7)	
Stage	Stage II + III	119 (63.3)	33 (16.9)	<0.001
	Stage IV	69 (36.7)	162 (83.1)	
Location	Cervical	50 (26.6)	5 (2.6)	<0.001
	Upper	54 (28.7)	37 (19.0)	
	Mid	41 (21.8)	68 (34.9)	
	Lower	43 (22.9)	85 (43.6)	
Histology	Squamous cell carcinoma	185 (98.4)	181 (92.8)	0.087
	Others	3 (1.6)	14 (7.2)	
Length of obstruction	cm	6.72 ± 3.17	6.53 ± 2.70	0.85
Chemotherapy	None	27 (14.4)	49 (25.1)	<0.001
	Before procedure	71 (37.8)	100 (51.3)	
	After procedure	90 (47.9)	46 (23.6)	
Radiotherapy	None	46 (24.5)	115 (59.0)	<0.001
	Before procedure	60 (31.9)	42 (21.5)	
	After procedure	82 (43.6)	38 (19.5)	
Surgery	None	130 (69.1)	187 (95.9)	<0.001
	Before procedure	1 (0.5)	0 (0.0)	
	After procedure	57 (30.3)	8 (4.1)	

Table 2. Baseline characteristics of patients with malignant esophageal obstruction undergoing gastrostomy and stent insertion, after propensity score matching (94 matched pairs).

		Gastrostomy Group (*n* = 94)	Stent Group (*n* = 94)	SMD
Age	year	64.67 ± 9.66	65.36 ± 10.05	0.070
Gender	Male	84 (89.4)	86 (91.5)	0.072
	Female	10 (10.6)	8 (8.5)	
Stage	Stage II + III	37 (39.4)	33 (35.1)	0.088
	Stage IV	57 (60.6)	61 (64.9)	
Location	Cervical	8 (8.5)	5 (5.3)	−0.13
	Upper	29 (30.9)	28 (29.8)	−0.023
	Mid	28 (29.8)	28 (29.8)	0.00
	Lower	29 (30.9)	33 (35.1)	0.091
Histology	Squamous cell carcinoma	92 (97.9)	93 (98.9)	0.085
	Others	2 (2.1)	1 (1.1)	
Length of obstruction	cm	7.06 ± 3.32	6.75 ± 2.93	0.098
Chemotherapy	None	20 (21.3)	23 (24.5)	0.076
	Before procedure	42 (44.7)	40 (42.6)	−0.043
	After procedure	32 (34.0)	31 (33.0)	−0.023
Radiotherapy	None	40 (42.6)	42 (44.7)	0.043
	Before procedure	28 (29.8)	29 (30.9)	0.023
	After procedure	26 (27.7)	23 (24.5)	−0.073
Surgery	None	89 (94.7)	88 (93.6)	−0.045
	Before procedure	1 (1.1)	0 (0.0)	−0.15
	After procedure	4 (4.3)	6 (6.4)	0.095

SMD, standardized mean difference.

3.2. Primary Outcomes of the Propensity Score-Matched Cohort

Primary outcomes in the matched cohort are shown in Table 3. A total of 14 procedure-related adverse events occurred. In the stent group, there were 6 adverse events; stent broken ($n = 1$), stent migration ($n = 2$), stent induced tracheal compression ($n = 1$), tumor bleeding ($n = 1$), and chest pain ($n = 1$). In the gastrostomy group, there were 8 adverse events; leakage ($n = 1$), gastrostomy site infection ($n = 4$), and peritonitis ($n = 3$). The gastrostomy group showed less decrease in serum albumin level and needed less additional interventions (AIs) than the stent group after the procedure (p-value 0.021 and <0.001, respectively). In the stent group, AIs included stent reposition/removal due to migration, stent removal due to chest pain, and gastrostomy due to tumor ingrowth into stent. In the gastrostomy group, AIs included gastrostomy tube removal due to localized infection and gastrostomy revision due to leakage. The number of AIs was lower in the gastrostomy group than in the stent group (0.04 ± 0.25 vs. 0.31 ± 0.61, p-value < 0.001). There was no significant difference between the two groups with regard to body weight change, occurrence of procedure-related adverse events, and aspiration pneumonia.

3.3. Overall Survival of the Study Population

During median follow-up of 5 months (2.6–9.3), the gastrostomy group showed higher OS rates in the unmatched cohort ($n = 383$, Figure 2). However, after propensity score matching, there was no significant difference between the two groups (94 pairs) in OS (Figure 3).

3.4. Factors Associated with Overall Survival in Propensity Score-Matched Cohort

To adjust for potential confounders after propensity score matching, Cox proportional hazards regression model was conducted. Multivariable analysis showed that chemotherapy after procedure, surgery before or after procedure, and type of procedure (gastrostomy) were the independent factors associated with OS (Table 4).

Table 3. Comparison of primary outcomes between patients with malignant esophageal obstruction undergoing gastrostomy and stent insertion, in the propensity score matched cohort (94 matched pairs).

		Gastrostomy Group (n = 94)	Stent Group (n = 94)	p-Value
Weight change [1]	kg	−0.69 ±2.56	−0.27 ± 3.48	0.58
Albumin change [1]	g/dL	−0.15 ± 0.57	−0.41 ± 0.59	0.021
Additional intervention	None	92 (97.9)	72 (76.6)	<0.001
	Yes	2 (2.1)	22 (23.4)	
Number of additional interventions		0.04 ± 0.25	0.31 ± 0.61	<0.001
Procedure-related complications	None	90 (95.7)	91 (96.8)	1.00
	Yes	4 (4.3)	3 (3.2)	
Aspiration pneumonia	None	81 (86.2)	81 (86.2)	1.00
	Yes	13 (13.8)	13 (13.8)	

[1] After procedure.

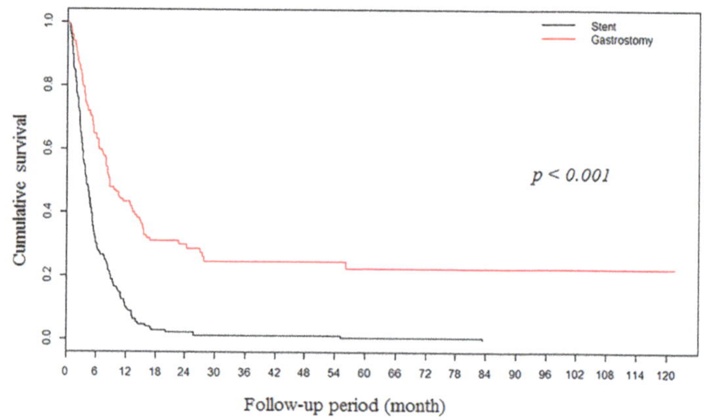

Figure 2. Kaplan–Meier survival curves showing overall survival rates of 383 patients with malignant esophageal obstruction after gastrostomy (n = 188) and stent insertion (n = 195).

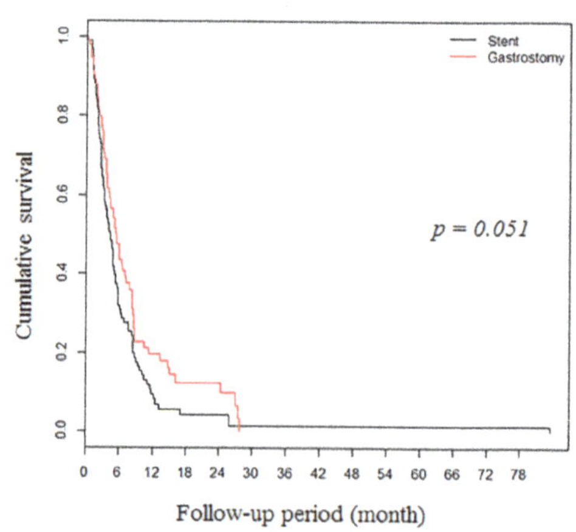

Figure 3. Kaplan–Meier survival curves showing overall survival rates in the propensity score-matched cohort (94 matched pairs) with patients with malignant esophageal obstruction after gastrostomy and stent insertion.

Table 4. Factors associated with overall survival in propensity score matched cohort with malignant esophageal obstruction (94 matched pairs).

		Multivariable Analysis	
		HR (95% CI)	*p*-Value
Stage	Stage II + III	1	
	Stage IV	1.43 (1.00–2.06)	0.052
Location	Cervical	1	0.30
	Upper	1.22 (0.64–2.34)	0.54
	Mid	1.63 (0.84–3.16)	0.15
	Lower	1.15 (0.59–2.23)	0.69
Chemotherapy	None	1	0.072
	Before procedure	0.89 (0.57–1.39)	0.62
	After procedure	0.60 (0.38–0.95)	0.029
Radiotherapy	None	1	0.031
	Before procedure	1.55 (1.00–2.39)	0.050
	After procedure	0.79 (0.51–1.22)	0.29
Surgery	None	1	<0.001
	Before procedure	123.08 (10.15–1492.60)	<0.001
	After procedure	0.34 (0.14–0.82)	0.016
Type of procedure	Stent	1	
	Gastrostomy	0.69 (0.50–0.95)	0.024

HR, hazard ratio; CI, confidence interval.

4. Discussion

Currently, patients with malignant esophageal obstruction are either subjected to gastrostomy or SEMS insertion. It is mainly because the maintenance of nutritional status through these procedures provides better clinical outcomes, including improvement of QoL and survival benefits [17,18]. However, it has not been established yet whether these are the optimal treatment strategies. In our study, we demonstrated that PG was superior to SEMS, by assessing the patients' nutritional status using propensity score matching analysis within a large cohort.

Systematic review and meta-analysis of nine studies comprising 180 patients showed that stent insertion relieved dysphagia immediately and led to an increase in both body weight and serum albumin level, but chest pain and stent migration occurred with high incidence rates (51.4% and 32%, respectively) [4]. A retrospective study with propensity score matching conducted by Mariette et al. (2015) found that SEMS insertion was a predictor of poor prognosis with adjusted confounding factors (hazard ratio (HR) 1.6, 95% confidence interval (CI) 1.02–2.5), even though the median time of recurrence and three-year survival were found to be reduced in the SEMS group [19]. Based on the above-mentioned two studies, the European Society of Gastrointestinal Endoscopy does not recommend the temporary placement of an SEMS/SEPS (self-expandable plastic stent) for malignant dysphagia as a bridge to surgery or before pre-operative chemoradiotherapy, and suggests other options, such as placement of a feeding tube (strong recommendation based on low-quality evidence) [3,13]. In our study, 93% (162/181) of patients in the stent group and 67% (129/188) in the PG group underwent procedure in palliative care. Because several cases were performed before current ESGE guidelines had been established, some patients underwent SEMS between late 1990 and early 2010 for non-palliative care. However, SEMS insertion was performed recently in our hospital only for palliative care, consistent with current ESGE guidelines. In our previous retrospective study, we first compared the two procedures, and then observed that PG feeding was associated with better OS in patients with malignant esophageal obstruction compared to SEMS insertion by stabilizing the nutritional status (HR 0.56, 95% CI 0.36–0.87) [20]. However, our previous study had several limitations that could be attributed to a relatively small sample size and risk of bias. To minimize the risk of bias, in the present study, we increased the sample size and analyzed the cohort thoroughly with propensity score matching and the obtained results were taken into consideration for further interpretation. Thereafter, we observed that gastrostomy resulted in less decrease in the serum albumin level compared to the stent group, after the procedure. This parameter reflected the nutritional status, and consequently, better outcomes. Indeed, in the present study, we showed that PG was an independent factor associated with OS (HR 0.69, 95% CI 0.50–0.95). Although there was no significant difference in OS between the two groups ($p = 0.051$), it was still very close to statistical significance. In contrast to other studies, there was no difference in procedure-related adverse outcomes, including chest pain, between the two groups. It could be explained by the assumption that at our hospital, PG was performed in cases of cervical malignant esophageal obstruction, which later resulted in chest pain due to stent insertion. Additionally, according to a recent prospective study conducted by Yu et al. (2018) comparing nasogastric tube, stent, and ostomy tube feeding, stent led to a poor QoL when compared with the other two groups [21]. On comprehensive evaluation, gastrostomy tended to be superior to SEMS in general aspect, including nutritional status and QoL.

In multivariable analysis, surgery after procedure showed the best OS, followed by only procedure without surgery, and surgery before procedure. Compared to patients who underwent only procedure without surgery, the risk of mortality was higher for those who underwent surgery before procedure (HR 123.08, 95% CI 10.15–1492.60). However, only one patient underwent surgery before procedure and the range of 95% CI was very wide. Considering these points, the risk was overestimated and it should be interpreted carefully.

Our study had some limitations. The first limitation is that it is a retrospective study. However, as many previous studies have demonstrated the potential advantages of gastrostomy, it was difficult to conduct a randomized controlled trial (RCT) in an ethical manner. Additionally, conducting prospective RCT to compare SEMS insertion and gastrostomy for the treatment of malignant esophageal obstruction was practically impossible. Instead, to obtain a high level of evidence close to a RCT, we first analyzed the study population thoroughly with propensity score matching, and then interpreted the results that were obtained. Second, because of the limitation of conducting a retrospective study, QoL was not assessed, which was recently identified as an important part of palliative treatment in patients with malignant esophageal obstruction [10]. Instead of error-prone symptom evaluation by retrospective analysis, we decided to assess objective parameters, such as body weight and serum albumin levels, as primary outcomes. Third, nutritional status was determined one month after the procedure, which might be too short a time for decreased body stores to be evidenced by weight loss and a decreased albumin level. Also albumin, an acute phase reactant, could not be used as a good nutritional marker. However, due to high mortality and poor general condition related to the high risk of malnutrition, there were a considerable number of missing data regarding body weight and albumin, more than two months after procedure; yet several studies have used albumin inevitably as a nutritional marker. Unavoidably, we assessed procedural nutritional status as a change of serum albumin level and body weight between baseline and one month after procedure. Fourth, esophageal squamous cell carcinoma is more common in Asian than Western populations, and only three esophageal adenocarcinoma cases were included in our study. So, this study may be unclear when generalizing the results to Western patients with malignant esophageal obstruction. Despite these limitations, we believe that we have provided enough evidence to support the advantage of PG through a well-designed analysis using propensity score matching.

5. Conclusions

In conclusion, we suggest that gastrostomy may be preferred over stent insertion in patients with malignant esophageal obstruction, considering the nutritional and survival benefits.

Author Contributions: Conceptualization, J.H.S., and Y.W.M.; methodology, J.H.S., Y.W.M., and K.K.; software, J.H.S.; validation, J.H.S., and Y.W.M.; formal analysis, J.H.S., Y.W.M., and K.K.; investigation, J.H.S., and Y.W.M.; resources, J.H.S.; data curation, J.H.S., and J.K.; writing—original draft preparation, J.H.S.; writing—review and editing, Y.W.M.; visualization, J.H.S., and Y.W.M.; supervision, Y.W.M., H.L., B.-H.M., J.H.L., P.-L.R., and J.J.K. All authors have read and agreed to the published version of the manuscript.

Funding: This research received no external funding.

Acknowledgments: We would like to thank all the participants.

Conflicts of Interest: The authors declare no conflict of interest.

References

1. Ferlay, J.; Steliarova-Foucher, E.; Lortet-Tieulent, J.; Rosso, S.; Coebergh, J.W.; Comber, H.; Forman, D.; Bray, F. Cancer incidence and mortality patterns in Europe: Estimates for 40 countries in 2012. *Eur. J. Cancer* **2013**, *49*, 1374–1403. [CrossRef]
2. Bozzetti, F. Screening the nutritional status in oncology: A preliminary report on 1,000 outpatients. *Support Care Cancer* **2009**, *17*, 279–284. [CrossRef] [PubMed]
3. Spaander, M.C.; Baron, T.H.; Siersema, P.D.; Fuccio, L.; Schumacher, B.; Escorsell, A.; Garcia-Pagan, J.C.; Dumonceau, J.M.; Conio, M.; de Ceglie, A.; et al. Esophageal stenting for benign and malignant disease: European Society of Gastrointestinal Endoscopy (ESGE) Clinical Guideline. *Endoscopy* **2016**, *48*, 939–948. [CrossRef] [PubMed]
4. Nagaraja, V.; Cox, M.R.; Eslick, G.D. Safety and efficacy of esophageal stents preceding or during neoadjuvant chemotherapy for esophageal cancer: A systematic review and meta-analysis. *J. Gastrointest. Oncol.* **2014**, *5*, 119–126. [PubMed]
5. Lopes, T.L.; Eloubeidi, M.A. A pilot study of fully covered self-expandable metal stents prior to neoadjuvant therapy for locally advanced esophageal cancer. *Dis. Esophagus* **2010**, *23*, 309–315. [CrossRef] [PubMed]
6. Shenfine, J.; McNamee, P.; Steen, N.; Bond, J.; Griffin, S.M. A randomized controlled clinical trial of palliative therapies for patients with inoperable esophageal cancer. *Am. J. Gastroenterol.* **2009**, *104*, 1674–1685. [CrossRef] [PubMed]
7. Senft, M.; Fietkau, R.; Iro, H.; Sailer, D.; Sauer, R. The influence of supportive nutritional therapy via percutaneous endoscopically guided gastrostomy on the quality of life of cancer patients. *Support Care Cancer* **1993**, *1*, 272–275. [CrossRef]
8. Bower, M.R.; Martin, R.C., 2nd. Nutritional management during neoadjuvant therapy for esophageal cancer. *J. Surg. Oncol.* **2009**, *100*, 82–87. [CrossRef]
9. Loser, C.; Aschl, G.; Hebuterne, X.; Mathus-Vliegen, E.M.; Muscaritoli, M.; Niv, Y.; Rollins, H.; Singer, P.; Skelly, R.H. ESPEN guidelines on artificial enteral nutrition–percutaneous endoscopic gastrostomy (PEG). *Clin. Nutr.* **2005**, *24*, 848–861.
10. Sofue, K.; Takeuchi, Y.; Tsurusaki, M.; Shibamoto, K.; Sakamoto, N.; Kitajima, K.; Sone, M.; Sugimura, K.; Arai, Y. Value of percutaneous radiologic gastrostomy for patients with advanced esophageal cancer. *Ann. Surg. Oncol.* **2016**, *23*, 3623–3631. [CrossRef]
11. Zuercher, B.F.; Grosjean, P.; Monnier, P. Percutaneous endoscopic gastrostomy in head and neck cancer patients: Indications, techniques, complications and results. *Eur. Arch. Otorhinolaryngol.* **2011**, *268*, 623–629. [CrossRef] [PubMed]
12. Martin, R.C., 2nd; Cannon, R.M.; Brown, R.E.; Ellis, S.F.; Williams, S.; Scoggins, C.R.; Abbas, A.E. Evaluation of quality of life following placement of self-expanding plastic stents as a bridge to surgery in patients receiving neoadjuvant therapy for esophageal cancer. *Oncologist* **2014**, *19*, 259–265. [CrossRef] [PubMed]
13. De la Mora Levy, J.G.; Manzano-Robleda, M.D.C. Nutrition before chemoradiotherapy or surgery: Temporary esophageal stents or tube feeding? Is the evidence hard to swallow? *Gastrointest. Endosc.* **2018**, *88*, 32–34. [CrossRef] [PubMed]
14. Rice, T.W.; Blackstone, E.H.; Rusch, V.W. 7th edition of the AJCC cancer staging manual: Esophagus and esophagogastric junction. *Ann. Surg. Oncol.* **2010**, *17*, 1721–1724. [CrossRef]
15. Ho, D.; Imai, K.; King, G.; Stuart, E.A. Matching as nonparametric preprocessing for reducing model dependence in parametric causal inference. *Political Anal.* **2007**, *15*, 199–236. [CrossRef]
16. Stuart, E.A. Matching methods for causal inference: A review and a look forward. *Stat. Sci. A Rev. J. Inst. Math. Stat.* **2010**, *25*, 1. [CrossRef]
17. Anandavadivelan, P.; Lagergren, P. Cachexia in patients with oesophageal cancer. *Nat. Rev. Clin. Oncol.* **2016**, *13*, 185–198. [CrossRef]
18. Huddy, J.R.; Huddy, F.M.S.; Markar, S.R.; Tucker, O. Nutritional optimization during neoadjuvant therapy prior to surgical resection of esophageal cancer-a narrative review. *Dis. Esophagus* **2018**, *31*, 1–11. [CrossRef]

19. Mariette, C.; Gronnier, C.; Duhamel, A.; Mabrut, J.Y.; Bail, J.P.; Carrere, N.; Lefevre, J.H.; Meunier, B.; Collet, D.; Piessen, G. Self-expanding covered metallic stent as a bridge to surgery in esophageal cancer: Impact on oncologic outcomes. *J. Am. Coll. Surg.* **2015**, *220*, 287–296. [CrossRef]
20. Min, Y.W.; Jang, E.Y.; Jung, J.H.; Lee, H.; Min, B.H.; Lee, J.H.; Rhee, P.L.; Kim, J.J. Comparison between gastrostomy feeding and self-expandable metal stent insertion for patients with esophageal cancer and dysphagia. *PLoS ONE* **2017**, *12*, e0179522. [CrossRef]
21. Yu, F.J.; Shih, H.Y.; Wu, C.Y.; Chuang, Y.S.; Lee, J.Y.; Li, H.P.; Fang, P.T.; Tsai, D.L.; Chou, S.H.; Wu, I.C. Enteral nutrition and quality of life in patients undergoing chemoradiotherapy for esophageal carcinoma: A comparison of nasogastric tube, esophageal stent, and ostomy tube feeding. *Gastrointest. Endosc.* **2018**, *88*, 21–31. [CrossRef] [PubMed]

nutrients

MDPI

Review

Feasibility, Process, and Effects of Short-Term Calorie Reduction in Cancer Patients Receiving Chemotherapy: An Integrative Review

Chia-Chun Tang [1,*], Hsi Chen [1], Tai-Chung Huang [2], Wei-Wen Wu [1], Jing-Mei Lin [3] and Feng-Ming Tien [2]

1 School of Nursing, College of Medicine, National Taiwan University, Taipei 10617, Taiwan; hchsichen@ntu.edu.tw (H.C.); weiwen@ntu.edu.tw (W.-W.W.)
2 Division of Hematology, Department of Internal Medicine, National Taiwan University Hospital, Taipei 10617, Taiwan; tch01@ntu.edu.tw (T.-C.H.); b92401007@ntu.edu.tw (F.-M.T.)
3 Department of Dietetics, National Taiwan University Hospital, Taipei 10617, Taiwan; kyomilin@ntuh.gov.tw
* Correspondence: chiatang@ntu.edu.tw; Tel.: +886-2-2312-3456 (ext. 88436)

Received: 14 August 2020; Accepted: 11 September 2020; Published: 15 September 2020

Abstract: Recent preclinical studies have shown the potential benefits of short-term calorie reduction (SCR) on cancer treatment. In this integrative review, we aimed to identify and synthesize current evidence regarding the feasibility, process, and effects of SCR in cancer patients receiving chemotherapy. PubMed, Cumulative Index to Nursing and Allied Health Literature, Ovid Medline, PsychINFO, and Embase were searched for original research articles using various combinations of Medical Subject Heading terms. Among the 311 articles identified, seven studies met the inclusion criteria. The majority of the reviewed studies were small randomized controlled trials or cohort study with fair quality. The results suggest that SCR is safe and feasible. SCR is typically arranged around the chemotherapy, with the duration ranging from 24 to 96 h. Most studies examined the protective effects of SCR on normal cells during chemotherapy. The evidence supports that SCR had the potential to enhance both the physical and psychological wellbeing of patients during chemotherapy. SCR is a cost-effective intervention with great potential. Future well-controlled studies with sufficient sample sizes are needed to examine the full and long-term effects of SCR and its mechanism of action.

Keywords: integrative review; short-term calorie reduction; fasting; cancer; chemotherapy; calorie restriction

1. Introduction

Emerging evidence has shown that glucose and caloric intake have powerful impacts on health, in both the general and the critically ill population, including cancer patients [1–4]. High glucose levels can contribute to a vicious circle that affects cancer formation, treatment, and progression [5,6]. Recent expert opinions suggest that glucose reduction and calorie control could enhance cancer treatments and improve patient outcomes [7,8]. There are at least four proposed mechanisms of how calorie restriction (CR), or fasting, affects tumor growth and treatment effectiveness. First, CR increases tumor cells' sensitivity to anticancer therapy by promoting apoptosis within tumors, which reduces levels of growth factors such as insulin-like growth factor-1 (IGF-1), and by inducing autophagy via the activation of AMP-activated protein (AMPK)/the mammalian target of rapamycin (mTOR) pathway. Second, in contrast, CR selectively protects normal cells from stress and toxicity of anticancer therapy because they react oppositely to the aforementioned interferences. Moreover, CR-induced autophagy may promote tissue regeneration. Third, by decreasing inflammation and increasing circulating T cells, CR establishes an environment that is unfavorable to tumor growth. Fourth, CR inhibits tumor growth by reducing the expression of factors that promote neovascularization of tumors [8–13].

Compared with a chronic 20–40% CR, which requires weeks to months to detect its effects on cancer progression, a short-term CR (SCR; for example, a calorie reduction of over 50% lasting no longer than a week) has shown immediate effects on enhancing the therapeutic effects of chemotherapy and protecting normal cells from drug toxicity [12,14,15]. SCR also seems to be safe, and does not cause weight loss, which is the main side effect of chronic CR [10]. Several in vivo (mouse models) and in vitro studies have demonstrated positive effects of SCR on suppressing tumor growth (for example, in pancreatic cancer and hepatocellular cancer) and enhancing the effects of chemotherapeutic agents (such as, doxorubicin, gemcitabine, and sorafenib). The in vivo studies have shown that SCR significantly increases chemotherapy effects by inhibiting tumor growth, cellular proliferation, and metabolism [12,14,15]. D'Aronzo and colleagues even demonstrated that SCR alone is just as effective as SCR plus gemicitabine in inhibiting pancreatic cancer cell migration in vitro and using animal models [12]. Some evidence has indicated that undertaking SCR (fasting for 24–72 h with access to water or eating a diet that mimics fasting) prior to chemotherapy protects normal cells, regulates glycemia, and enhances the therapeutic effects of chemotherapy [12,15,16]. Di Biase and colleagues found that SCR decreased doxorubicin-induced cardiotoxicity and prevented hyperglycemia in mice, thereby providing protection from glucose- and dexamethasone-dependent sensitization to doxorubicin [16].

Although the results from animal studies are promising and human trials have begun, clinical oncologists to date only provide universal and generic dietary guidelines to all cancer patients [17]. For example, in the latest nutrition guide published by the American Institute for Cancer Research, Livestrong Foundation, and Savor Health [18], the main nutrition recommendation for all cancer patients under treatment is to eat a healthy and clean diet. The European Society for Clinical Nutrition and Metabolism (ESPEN) guideline for patients undergoing drug treatment is to "ensure adequate nutritional intake" [19]. Several experts have pointed out that the level of evidence for these recommendations is low [17,19]. In fact, to our knowledge, no nutrition guidelines or recommendations have ever mentioned any form of SCR. This may be due to the early stage of clinical studies and the lack of systematic reviews that evaluate and synthesize current SCR evidence. The vague recommendation is insufficient to answer the necessary but unanswered question of "how to eat right?" In a survey ($n = 1335$), more than two thirds of the patients with cancer indicated that they had questions regarding nutrition or food intake [20]. In contrast, a considerable number of cancer patients (39–76%) have reported unmet needs regarding nutrition-related information or issues [21,22]. Therefore, the aim of this review is to identify and synthesize current evidence regarding the feasibility, process, and effects of SCR in cancer patients receiving chemotherapy. The findings from this review will identify areas for future research, aid in reexamining nutrition guidelines and enhance evidence-based clinical practice.

2. Materials and Methods

It is important to analyze all the available data for new concepts and underexplored research areas such as SCR. Therefore, the method of integrative review was selected; this allowed us to include as much evidence as possible, regardless of the study design and type of data. We followed the well-established review process described by Whittemore and Knafl, which included the following: problem identification, literature search, data evaluation and analysis, and presentation of the results [23].

2.1. Literature Search

We searched the following five databases for articles describing SCR in cancer patients undergoing chemotherapy: PubMed, CINAHL, Ovid Medline, PsychINFO, and Embase. Several combinations of Medical Subject Heading (MeSH) terms were used in different databases (Table 1). The original studies exploring the effects of SCR on cancer patients receiving chemotherapy were included only if they were written in English, included human cancer patients, and were peer-reviewed. We did not set any limits on the dates of publication and the final date of the search is the 6 August 2020. Articles were excluded if they did not meet any one of the aforementioned criteria or if they focused on the effects of food on drug pharmacokinetics. The eligibility of the literature was determined by screening the titles, then the abstracts, finally, a full-text review. In addition, the reference lists of each included article and the website ClinicalTrials.gov were searched to identify relevant studies. EndNote X8 was then used to sort citations and remove duplicates.

Table 1. Searched databases, searching strategies, and the number of initial results.

Databases	Searching Strategies: Combination of Medical Subheadings	Initial Results
PubMed	("fasting" OR "calorie restricted") AND "chemotherapy"	238
Ovid Medline	("fasting" OR "diet, carbohydrate-restricted" OR "calorie restriction") AND ("maintenance chemotherapy" OR "induction chemotherapy" OR "consolidation chemotherapy" OR "chemotherapy, adjuvant" OR "chemotherapy, cancer, regional perfusion")	9
CINAHL	("fasting" OR ("preprocedural Fasting" OR "restricted diet" OR "diet, reducing" OR "diet, low carbohydrate") AND ("chemotherapy, cancer" OR "chemotherapy, adjuvant" OR "chemotherapy care (Saba CCC)" OR "chemotherapy management (Iowa NIC)" OR "antineoplastic agents, combined")	7
PsychINFO	("calories" OR "dietary restraint") AND "chemotherapy"	38
Embase	"caloric restriction" AND "cancer chemotherapy"	19

2.2. Data Evaluation and Analysis

We fully reviewed and rated the included literature in terms of its level of evidence and level of quality presented, which reflects the generalizability of a study. The definition of each level of evidence are presented in Table 2 which was modified from Wright and colleagues [24]. There are four level of research quality: good (the risk of bias is very low and the results are considered to be valid), fair (the study is susceptible to some bias deemed not sufficient to invalidate its results), poor (there is a significant risk of bias), and not to be analyzed (there is a fatal flaw) [25,26]. Because the designs of the included studies vary, we employed four scales to evaluate the quality of the studies. Quantitative studies were evaluated on the basis of Quality Assessment Tools developed by methodologists from the NHLBI and Research Triangle Institute International. Specifically, the Quality Assessment of Controlled Intervention Studies [27] was used to evaluate randomized controlled trials, the Quality Assessment Tool for Observational Cohort and Cross-Sectional Studies [28] was chosen to assess prospective cohort studies, and the Quality Assessment Tool for Case Series Studies [27] was selected for case studies. Instructions for Evaluating Qualitative Literature [26] were employed for qualitative studies. Studies that met 75–100% criteria were determined to be of good quality while 50–74% criteria met signified a fair quality and 25–49% criteria met indicated poor quality. Next, study information was collected and categorized in a data collection file prepared by C.T. using Microsoft Word. Specifically, three kinds of information were collected: study characteristics (design, population, fasting plan, and type of chemotherapy), type of outcome measurements, and main study outcomes. All research activities were independently performed by C.T. and H.C. In case of discordant opinions, the research team discussed and solved these issues in regular meetings.

Table 2. Level of Evidence.

Level	Definition
I	Randomized controlled trial
II	Prospective cohort study or Poor-quality randomized controlled trial
III	Case-control study or Retrospective cohort study
IV	Case series
V	Expert opinion

Note. Modified from Wright, Swiontkowski, and Heckman (2003) [24].

3. Results

Initially, 311 articles were identified. After removing duplicates (*n* = 3), 308 articles were screened by title, which resulted in a total of 67 articles for abstract screening. Using the established criteria, 60 articles were excluded. Among the 60 articles, 60% (*n* = 36) were not complete original research articles; 33% (*n* = 20) presented irrelevant content; 5% (*n* = 3) did not include human samples; and 2% (*n* = 1) were not written in English. The remaining seven studies that were retained for full-text review were all included in the analysis (Figure 1).

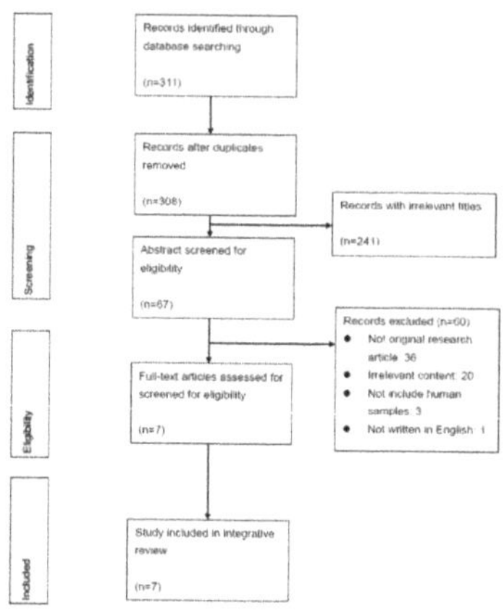

Figure 1. PRISMA Diagram of Search Results and Screening Process.

3.1. *Study Characteristics*

Among the seven studies included in this review (Table 3), one is a qualitative study and others have a quantitative design, including a case study, a cohort study, and randomized controlled trials (RCT, n = 4). The sample sizes ranged from 13 to 129. Five studies focused on gynecologic cancer populations [29–33] and the other two involved various types of cancer. In the five studies that stipulated strict timelines for SCR, the total period ranged between 24 and 96 h, with SCR typically starting 24–72 h before the chemotherapy and lasting for about 24 h after the completion of chemotherapy [29,30,32–34]. The other two studies observed participants' self-determined reduction practices, and thus presented large variations in the SCR timeframe—the patients started SCR 24–140 h prior to chemotherapy and ended it 5–56 h following chemotherapy [31,35]. The actual number of calories consumed during the practice of SCR differed across studies. Most studies required the participants to fast, allowing only non-caloric beverages. One study offered a rescue option to consume less than 200 kcal a day if fasting symptoms became apparent [34]. Bauersfeld et al. set the daily maximum total intake at 350 kcal [29] and de Groot et al. [32] designed a fasting mimicking diet with decreasing calorie amount over three days (200–1200 kcal). On the other hand, Zorn et al., instructed a group of patients to consume a 6-day normocaloric ketogenic diet before water fast. While a case study mentioned that some of their participants ate nothing except for water and vitamins [35], other studies did not specify if any nutritional supplements were used. The participants received various types of chemotherapy drugs and regimens, including taxanes, platinum, alkylating, anthracycline, antimetabolites, and IgG1 antibody. In terms of the level of evidence of the quantitative studies, the majority was level II small RCTs or cohort study (*n* = 4); others were level I RCT (*n* = 1) and level IV case series (*n* = 1) [24]. Using the aforementioned quality scales to evaluate, more than half of the studies had fair or poor quality (Tables 4 and 5). Only two studies were of good quality, including one RCT and one qualitative study (data not shown in table) [31,32]. The most obvious threats to the quality of RCT studies were the high drop-out rates and low adherence.

Table 3. Information of Reviewed Articles: Type of Design, Method, and Maine Results.

First Author, Year (Country)	Goal: Research Design	Sample Size, Population, Exclusion Criteria	Calorie Reduction Plan	Chemotherapy Regimen	Measuring Time and Outcome Measurements	Main Results
Dorff, 2016 (USA)	Determine the safety/feasibility of fasting prior to C/T: Cohort study (24/48/72 h.)	- $n = 20$ (6–7 patient/cohort) - various cancer types/stages - Exclusion: DM, BM ≤ 20.5, recent BW loss > 10 kg	**Dose/time**: escalating fast, up to 72 h (24 h before C/T completion → 48 h before C/T completion → if safe/feasible, then continue with 72 h (48 h before and 24 h after); if not, then try 48 h with specific low-calorie diet (repeat for at least 2 C/T cycles)) **Content**: NPO except for water and non-caloric beverage and rescue (<200 kcal/24 h. if fasting symptoms present)	- ≤2 days of Platinum-based combination C/T without concurrent radiation - May have begun C/T but still have 2 or more cycles - Standard antiemetic	○ Measuring before C/T, after fast, and 24 h after C/T completion: - Nutrition and metabolism status: Prealbumin, Insulin, Glucose, Ketones (β-hydroxybutyrate) - Side effects and fasting-related toxicities (CTCAE v4.0) - Hematological function - Endocrine parameters: IGF-1, IGFBPs - DNA damage: peripheral blood mononuclear cells - Treatment outcome: pathologic responses	○ Safety and tolerance of SCR: safe and feasible - Reasons for non-compliance: forget, social constraints, change of C/T plan, fail to regain weight - Nutrition and metabolism status *: β-hydroxybutyrate decreased in the 24-h group while it increased in 48- and 72-h groups ○ Main effects of SCR: - Effects on side effects, symptoms, and QOL *: decreasing C/T-related toxicity (nausea and vomiting) in all groups - Effects on hematological function: Insignificant trend of decreasing C/T-related grade 3–4 neutropenia - Effects on endocrine parameters: decreased but not-significant trend of IGF-1 - Effects on DNA damage in healthy cells *: mitigated in subjects who fasted for ≥48 h. - Effects of the treatment: no effects
Safdie, 2009 (USA)	Examine the safety of fasting before and after chemotherapy: Case study	- $n = 10$ - Various cancer types/stages	(Vary by cases) **Does/time**: 48–140 h prior to and/or 5–56 h following C/T (self-selected C/T cycles) **Content**: Some NPO except for water and vitamin, others unspecified **Control**: self-control	Individualized	○ At unspecified time points: - Self-reported symptoms: fatigue, weakness, hair loss, headaches, nausea, vomiting, diarrhea, abdominal cramps, mouth sores, dry mouth, short-term memory impairment, numbness, tingling, neuropathy motor - Hematological function: WBC, ANC, platelets - Treatment outcomes: CT-PET scan (one case)	○ Safety and tolerance of SCR: well-tolerated - Side effects: slight dizziness, hunger, and headache which did not interfere with daily activities. - Nutrition and metabolism status: weight loss was about 6–7 pounds which were regained quickly after resuming normal diet ○ Main effects of SCR: - Effects on side effects, symptoms, and QOL: self-reported reduction in multiple chemotherapy-induced side effects - Effects on hematological function: better recovery of blood counts, including less severe or shorter nadir of WBC/ANC/platelets - Effects of the treatment: better response to C/T in one patient

Table 3. *Cont.*

First Author, Year (Country)	Goal: Research Design	Sample Size, Population, Exclusion Criteria	Calorie Reduction Plan	Chemotherapy Regimen	Measuring Time and Outcome Measurements	Main Results
Bauersfeld, 2018 (Germany)	Examine the feasibility and effects of QOL of short-term fasting during C/T: Randomized, individually controlled trial	- $n = 34$ - Breast/ovarian cancer - Exclusion: BMI < 19, WHO performance status > 2, life expectancy < 3 months, DM, MI, stroke or pulmonary embolism within 3 months, unstable heart disease, renal failure, eating disorder, dementia, psychosis, impaired physical mobility	**Dose/time**: 60 h (36 h before and 24 h after C/T) **Content**: Unrestricted amounts of water, herbal tea, 2 × 100 cL vegetable juice and small standardizes quantities of light vegetable broth with a maximum total energy intake of 350 kcal/day **Control group**: self-controlled (group A: fast for the first half of C/T cycles (2 or 3 cycles) followed by normal diet); group B: vice versa sequence)	- 4–6 cycles of C/T: Taxanes, Platinum, Alkylating, Anthracycline, Antimetabolites, IgG1 antibody - Standard antiemetics and medication: dexamethasone and 5HT3 inhibitors	○ Baseline and 8 days after each C/T cycle: - Side effects and fasting-related toxicities: FACIT-G, FACIT-F ○ During and at the end of fasting: - Adverse events	○ Safety and tolerance of SCR: safe and well tolerated. - Reasons for non-compliance: headache, hyperventilation, weakness, aversion to fasting nutrition (n = 5, 10%) - Side effects: headache, hunger, nausea after intake of broth or juices, and orthostatic reaction; all were of low grade which did not interfere with daily activities - Nutrition and metabolism status: no significant changes in weight - More than 80% participants agreed that the fast was effective and wanted to continue the practice during C/T ○ Main effects of SCR: - Effects on side effects, symptoms, and QOL: less compromised QOL and reduced fatigue (Group A demonstrated a statistically significant and clinically meaningful benefits of fast on QOL and fatigue, while Group B only show clinically meaningful difference of the positive effect on QOL for fast intervention)
de Groot, 2015 (The Netherlands)	Identify the effects of 48 h fasting on C/T, including side effects, hematological parameters in breast cancer patient receiving TAC: Randomized controlled trial	- $n = 13$ - Stage II/III breast cancer - Exclusion: BMI < 19, WHO performance status > 2, life expectancy < 3 months, adequate function of bone marrow, liver, renal, and heart, DM	**Dose/time**: 48 h fasting (24 h before and after starting C/T) **Content**: NPO except for water or coffee/tea without sugar **Control group**: Eat according to the guidelines for healthy nutrition (n = 6, minimum of 2 pieces of fruit per day)	○ 6 cycles of (neo)-adjuvant TAC (docetaxel/doxorubicin/cyclophosphamide) ○ Anti-emetic agent and medication: dexamethasone, 5-HT3 receptor antagonist	○ Baseline (2 weeks before C/T), day 0 (prior to C/T) plus 30 min after C/T completion and day 7 of administration (only for hematological function, CRP, and DNA damage): - Nutrition and metabolic status: insulin, glucose - Hematological function: erythrocyte-, thrombocytes-, leukocyte count - DNA damage: γ-H2AX - Endocrine parameters: IGF-1, IGFBP3, TSH, triiodothyronine, free thyroxine - Inflammatory response: CRP ○ During C/T: self-reported side effects and CTCAE	○ Safety and tolerance of SCR: Participates were motivated to fast and the fast was well-tolerated and safe - Reasons of non-compliance: 2 withdraw at the 3rd cycle of C/T due to non-fasting-related signs (i.e., pyrosis and recurrent febrile neutropenia) - Nutrition and metabolism status: no significant changes ○ Main effects of SCR: - Effects on side effects, symptoms, and QOL: no significant effects - Effects on hematological function *: protect from C/T-related toxicity - Effects on endocrine parameters and inflammatory response: not significant - Effects on DNA damage in healthy cells (lymphocytes and myeloid cells) *: protect and promote recovery

Table 3. *Cont.*

First Author, Year (Country)	Goal: Research Design	Sample Size, Population, Exclusion Criteria	Calorie Reduction Plan	Chemotherapy Regimen	Measuring Time and Outcome Measurements	Main Results
de Groot, 2020 (The Netherlands)	Evaluate the impact of FMD on toxicity as well as on the radiological and pathological response to chemotherapy for breast cancer: Randomized controlled trial	- $n = 129$ - HER-2 (-), stage II/III breast cancer with WHO performance stage 0–2, BMI < 19 kg/m^2 - Exclusion: DM, allergies to designed food content, function impairment (bone marrow reserve, liver, renal, cardiac)	**Dose/time**: 4-day plant-based low amino-acid substitution diet (FMD, 3 days prior to and on the day of C/T) **Content**: decreased calorie intake from FMD (1200 kcal at day 1, 100 kcal at day 2–4) **Control group**: regular diet	○ 8 cycles of (neo)-adjuvant (docetaxel/doxorubicin/cyclophosphamide) or 6 cycles of (neo)-adjuvant FEC-T (5-fluorouracil, epirubicin, cyclophosphamide) ○ Dexamethasone was given before C/T only for control group	○ Baseline (day-1 or 0 of each chemotherapy): - Nutrition and metabolic status: insulin, glucose, ketone - Endocrine parameters: IGF-1 ○ Baseline (day-1 or 0 of the first cycle of chemotherapy) & 3 h after start of C/T: DNA damage: γ-H2AX ○ Baseline, halfway, at the end of therapy, and 6-month follow-up: EORTC QLQ-C30, ○ During C/T: CTCAE v4.03 ○ Halfway and at the end of the therapy: pCR ○ Halfway, at the end of the therapy, and 6-month follow-up: distress thermometer	○ Safety and tolerance of SCR: FMD was well-tolerated and safe - Reasons of non-compliance: the compliance decreased along with the C/T cycles (81.5% to 20% from cycle 1 to 8). The main reason of non-compliance was dislike of distinct components of the diet - Side-effects: no differences in toxicity - Nutrition and metabolism status *: lower insulin and glucose, ketones in urine (+) ○ Main effects of SCR: - Effects on side effects, symptoms, and QOL: not significant - Effects on endocrine parameters * and inflammatory response: lower IGF-1 - Effects on DNA damage in healthy cells *: protect and promote recovery - Pathological response *: better pCR
Zorn, 2020 (German)	Evaluate the influence of 96-h fasting on chemotherapy-induced toxicities in patients with gynecological cancer: controlled cross-over trial	- $n = 30$ - Gynecological cancer - Exclusion: malnutrition, eating disorders, DM, gout, severe cardiovascular disease, pregnancy or lactation, parental nutrition, administration of steroids or IGF-1 receptor blockers	**Dose/time**: 96 h fasting (72 h before and 24 h after starting C/T) or 6-day normocaloric ketogenic diet plus 96 h fasting **Content**: 25% of daily calorie requirement (400–600 kcal/day) with macronutrients revealed to a ketogenic composition **Control group**: everyone served as their own controls (2–3 cycles of SCR and 2–3 cycles of normal diet)	○ Paclitaxel/carboplatin ○ Epirubicin/cyclophosphamide ○ Docetaxel/cyclophosphamide	○ Baseline (before C/T), at each C/T, 3 weeks after the final C/T cycle: - Side effects and fasting-related toxicities: self-reported, CTCAE v4.0, EORTC QLQ-C30, EORTC QLQ-CIPN20, FACIT-Fatigue v 4.0 - Nutrition and metabolic status: insulin, body composition - Hematological function: erythrocyte-, thrombocytes-, leukocyte count - Endocrine parameters: IGF-1, TSH, triiodothyronine, free thyroxine - Inflammatory response: CRP	○ Safety and tolerance of SCR: well-tolerated and safe - Reasons for non-compliance: 2 withdraw because of fating-related discomfort and 19 withdraw due to non-fasting-related reasons - Side-effects: hunger, dizziness, weakness, and headache were mild - Nutrition and metabolism status *: reduction in insulin, BIA fat mass, weight (<5%), mean BIA cell mass, mean BIA phase angle; increase in BIA extracellular cell mass ○ Main effects of SCR: - Effects on side effects, symptoms, and QOL *: decreased symptoms such as stomatitis, headache, weakness, and overall symptom severity - Effects on hematological function *: decreased MCV and MCH - Effects on endocrine parameters and inflammatory response *: reduction in IGF-1, triiodothyronine; increase in free thyroxine

Table 3. *Cont.*

First Author, Year (Country)	Goal: Research Design	Sample Size, Population, Exclusion Criteria	Calorie Reduction Plan	Chemotherapy Regimen	Measuring Time and Outcome Measurements	Main Results
Mas, 2019 (France)	Explore the motivations to fast among cancer patients: Qualitative study	- $n = 16$ - Breast cancer	**Dose/time**: Having performed at least one 24 h fast before C/T within a year (duration ranges from a day and half to 7 days; C/T cycles range from one to 10 months). **Content**: not specified	Not mentioned	- Qualitative description of reason to fast, modalities of the fast, experience of fasting, related social support, barriers and facilitators of fasting - Satisfaction	○ Safety and tolerance of SCR: Patients believed that fasting is an efficacious non-conventional medicine that helps to reduce side effects of C/T for breast cancer. Patients expressed high level of satisfaction toward fasting. ○ Main effects of SCR: - Six themes emerge: 1. Main reason to fast: to lower the negative side effects of C/T, to regain control/act proactively during treatment (thus reduce the feelings of uncertainty and anxiety), improve C/T efficacy 2. Alternative authorities to the oncologist: conventional health care professions and other cancer patients' experience of fasting 3. Adapting the fast to social and lifestyle constraints: fasts were always performed with C/T 4. Fasting effects felt during chemotherapy: most of the patients reported positive physiological effects (especially nausea and vomiting) and about half experience psychological benefits 5. Barriers to (uncertainty of the effect of fasting, interference of meal sharing social life) and facilitators (anxiety regarding hospitalization, positive social support) of fasting during C/T 6. Seeking a more integrative medicine (although not supported by medical providers)

Note. 1. DM, Diabetes Mellitus; SCR, Shor-term Calorie Reduction; BMI, Body Mass Index; BW, Body Weight; kg, kilogram; kcal, kilocalorie; h, hour; n, size of sample; CTCAE, Common Terminology Criteria for Adverse Events; QoL, Quality Of Life; WBC, White Blood Cell; ANC, Absolute Neutrophil Counts; CT-PET, Computed Tomography-Positron Emission Tomography; MI, Myocardial Infarction; TAC, docetaxel/doxorubicin/cyclophosphamide; IGFBP3, Insulin Growth Factor Binding Protein 3; C/T, Chemotherapy; TSH, Thyroid-Stimulating Hormone; CRP, C-Reactive Protein; IGF-1, Insulin-like Growth Factor-1; IGFBPs, IGF Binding Proteins; NPO, Nothing by Mouth; cL, centilitre; FMD, fast mimic diet; pCR, pathological complete response; CTCAE, Common Terminology Criteria for Adverse Events; EORTC QLQ C30, The European Organization for Research and Treatment of Cancer quality of life questionnaire; EORTC QLQ CIPN20, The European Organization for Research and Treatment of Cancer quality of life Chemotherapy-Induced Peripheral Neuropathy; FACIT-F, the Functional Assessment of Chronic Illness Therapy-Fatigue; MCV, Mean corpuscular volume; MCH, Mean Corpuscular Hemoglobin. 2. * indicated that findings reached statistical or clinical significance.

Table 4. Quality and Evidence Level of Cohort Study and Case Report.

Quality Rating Criteria for Cohort Study	Study	Quality Rating Criteria for Case Report	Study
	Dorff, 2016		Safdie, 2009
Research question/objective was clearly stated	Yes	Research question/objective was clearly stated	No
Study population was clearly specified/defined	Yes	Study population was clearly specified/defined	Yes
Participation rate of eligible persons was ≥50%	Unclear	Cases were consecutive	No
Prespecified Inclusion/exclusion criteria	Yes	Subjects were comparable	No
Justification of sample size/power/variance/effect size	No	Intervention was clearly described	Yes
Exposure(s) measured prior to outcome(s) evaluation	Yes	Clearly defined, valid and reliable outcome measures	No
Sufficient timeframe to see a possible association	Yes	Adequate length of follow-up	Yes
Examine different exposure levels as related to the outcome	Yes	Well-described statistical methods	Not applicable
Clearly defined, valid and reliable exposure measures	Yes	Well-described results	Yes
Assessed the exposure(s) more than once over time	Yes		
Clearly defined valid and reliable outcome measures	Yes		
Outcome assessors were blinded to the exposure status of participants	Unclear		
Loss to follow-up was 20% or less	Yes		
Key potential confounding variables measured and adjusted statistically	No		
Suggesting Quality (% of criteria met)	Fair (71%)	Suggesting Quality (% of criteria met)	Fair (50%)
Level of Evidence	II	Level of Evidence	IV

Note. Level of quality was defined as: Good Quality (75–100% criteria met), Fair (50–74% criteria met), Poor (25–49% criteria met).

Table 5. Quality and Evidence Level of Randomized Controlled Trial.

Quality Rating Criteria	Studies			
	Bauersfeld, 2018	de Groot, 2015	de Groot, 2020	Zorn, 2020
Study was described as randomized or an RCT	Yes	Yes	Yes	No
Adequate randomization	Yes	Yes	Yes	No
Concealed treatment allocation	Yes	Unclear	Yes	Unclear
Study participants and providers were blinded to group assignment	Not applicable	Not applicable	Not applicable	Not applicable
People assessing the outcomes were blinded to the assignments	Unclear	Unclear	Yes	Unclear
Groups were similar at baseline on important characteristics	No	Yes	Yes	No
Overall drop-out rate at endpoint was ≤20% for treatment group	No	No	No	No
Differential drop-out rate between groups at endpoint was ≤15% or lower	Yes	Yes	No	No
Adherence to the intervention protocols were high	No	Yes	No	No
Other interventions were avoided or similar in the groups	Yes	Yes	Yes	Yes
Outcomes were assessed using valid and reliable measures	Yes	Yes	Yes	Yes
Sufficient sample size to be able to detect a difference with ≥80% power	Yes	No	Yes	Yes
Outcomes reported or subgroups analyzed were prespecified	Unclear	Unclear	Yes	Yes
All randomized participants were analyzed in the original group (intention-to-treat analysis)	Yes	Yes	Yes	Yes
Suggesting Level of Quality (% of criteria met)	Fair (62%)	Fair (62%)	Good (77%)	Poor (38%)
Level of Evidence	II	II	I	II

Note. Level of quality was defined as: Good Quality (75–100% criteria met), Fair (50–74% criteria met), Poor (25–49% criteria met).

3.2. Outcome Measurements

The following two categories of SCR outcomes were evaluated: safety/tolerance and overall effect. Specifically, the safety and tolerance of SCR were measured on the basis of the reasons for non-compliance with SCR, symptoms that were directly induced by SCR, and the change in nutrition or metabolism status. The effects of SCR were evaluated on the basis of its protective or regenerative effect on normal cells, ameliorative effect on inflammation, and sensitizing effect on tumor cells. The protective or regenerative effect on normal cells were evaluated on the basis of disease- or chemotherapy-associated side effects, quality of life, DNA damage in healthy cells, and hematological function. The reduction in inflammation was measured on the basis of the inflammatory response. The sensitizing effect on tumor cells to chemotherapy was evaluated using endocrine parameters and treatment outcomes. In addition to blood samples, several tools, such as Common Terminology Criteria for Adverse Events (CTCAE), the Functional Assessment of Chronic Illness Therapy-General (FACIT-G), The European Organization for Research and Treatment of Cancer quality of life questionnaires (EORTC

QLQ), and the Functional Assessment of Chronic Illness Therapy- Fatigue (FACIT-F) were employed to assess the side effects, symptoms, and quality of life. The researchers followed these variables across multiple cycles of chemotherapy in the following periods: "before each SCR and/or chemotherapy", "hours to days after each chemotherapy", "about a week after each chemotherapy", "at the end of chemotherapy treatment", and "6-month after treatment".

3.2.1. Safety and Tolerance of SCR

All studies concluded that SCR was safe, well-tolerated, and feasible [29–35]. More importantly, many participants expressed a strong motivation to undertake SCR and a desire to continue the practice in the future because of the perceived benefits of SCR, which included an increased sense of control [29–31].

The reported success rate of completing one cycle of SCR was above 80% [29,30]. However, the adherence decreased to below 50% when the researchers followed for more than three cycles [32,33]. Excluding non-SCR related symptoms (such as recurrent febrile neutropenia) and personal factors (such as forgetting, changing chemotherapy plan, and others), the reasons for withdrawal included headache, hyperventilation, weakness, failure to regain weight, aversion to fasting nutrition, and social constraints [29,30,32–35]. The qualitative study also reported social constraints as barriers to SCR—the patients who performed self-initiated SCR indicated that the protocol interfered with meal-sharing in their social lives. They also highlighted that the uncertainty surrounding the effects of fasting could be a barrier to SCR. In contrast, anxiety regarding hospitalization and positive social support might facilitate fasting behavior [31].

All researchers concluded that the possible side effects of SCR were mild, and that they either did not interfere with daily activities or did not require special treatment. The following side effects were noted: hunger, fatigue, dizziness, headache, hypoglycemia, weight loss, hyponatremia, orthostatic reaction or hypotension, and nausea after taking broth or juice [29,33–35]. Although weight loss may be an expected side effect of SCR, the studies showed that the loss of body weight was absent or minimal (about 6–7 pounds, <5%) [29,33,35], and that it was regained quickly after resuming a normal diet [29,35]. While pilot studies reported that no obvious changes in parameters related to nutrition and metabolism, such as prealbumin, insulin, and glucose, were observed [30,34], larger RCTs indicated that glucose and insulin were significantly lower in SCR groups before and during the treatment than controls [32,33]. The duration of fasting significantly affected ketone levels: de Groot and colleagues noted a decreasing trend in β-hydroxybutyrate levels (a type of ketone body) in 24-h fasting groups and an increasing trend in groups that fasted for more than 48 h [30]. The same research group later reported that ketone bodies were more likely to be positive in patients that performed SCR compared to regular diet [32]. In one study that examined body composition, the results showed decreased bioelectrical impedance analysis (BIA) fat mass, BIA body cell mass, mean BIA phase angle, and increased BIA extracellular cell mass [33].

3.2.2. Effects of SCR

The results of the six quantitative studies show mixed but overall positive findings regarding the effects of SCR. Most of the studies focused on SCR's protective or regenerative effect on normal cells, including chemotherapy-related side effects or symptoms [29,30,32–35], quality of life [29–33,35], hematological function [30,33–35], and DNA damage [30,32,34]. Five studies also examined endocrine parameters and/or treatment outcomes [30,32–35] to evaluate the sensitizing effects of tumor cells to chemotherapy. De Groot and colleagues measured inflammatory response [30].

SCR's Protective or Regenerative Effect on Normal Cells

Several studies suggested that SCR significantly reduces multiple chemotherapy-related side effects, such as nausea, vomiting, stomatitis, fatigue, headache, and overall symptom burden [29,33–35], and improved quality of life [29,35]. However, some did not find a significant reduction in side effects [30,32] or an improvement in the quality of life [30,32,33]. Zorn et al. pointed out a significant relationship between SCR and fewer chemotherapy postpones. The findings from a qualitative study that examined patients' motivation of self-initiated SCR reported that patients started SCR because they thought that it could mitigate the side effects of chemotherapy [31]. In fact, most of the patients reported positive physiological effects after fasting, and half of them experienced psychological benefits such as a reduction in feelings of uncertainty and anxiety [31].

To determine how SCR preserves or regenerates hematological function, the number and changes of erythrocytes, thrombocytes, and leukocytes were examined. All the studies that examined hematological function reported the protective effect of SCR [30,33–35], although the result from one study was insignificant [34]. Specifically, one study found that the erythrocyte and thrombocyte counts were significantly higher in the SCR group than in the control group one week or even 21 days after chemotherapy [30]. The results from another study showed a significantly milder neutropenia in patients who had fasted for longer than 48 h than in patients who had fasted for 24 h [34]. Zorn et al. (2020) found a significant decrease in mean corpuscular cell volume (MCV) and mean corpuscular hemoglobin (MCH).

Three studies looked at SCR's protective effect on chemotherapy-induced DNA damage, which was based on peripheral blood mononuclear cells. The results are encouraging [30,32,34]. Specifically, while DNA damage was obvious in all patients immediately after chemotherapy, patients who had fasted showed less chemotherapy-induced DNA damage 30 min to seven days later compared to the non-fasting group [30,32]. Further, one study that compared outcomes of 24-, 48-, and 72-h fasting specified that this protective effect was only observed in participants who had fasted for 48 h or longer [34].

Sensitizing Tumor Cells to Chemotherapy

As IGF-1, insulin-like growth factor-binding protein (IGFBPs), thyroid-stimulating hormone (TSH), triiodothyronine (fT3), and free thyroxine (fT4) were evaluated, a trend of decreasing IGF-1 [30,32–34], decreased fT3, and increased fT4 was found [33]. These indicators were measured at baseline, after fasting (but before chemotherapy) [30,34], and 24 h after chemotherapy [34]. In terms of pathological responses, the results from one study that involved a small group of patients showed no obvious impact of SCR on chemotherapy [34]. However, a large RCT showed that three times more partial or complete pathological responses were observed in patients performing SCR than in patients eating a regular diet [32]. From patients' perspectives, they indicated that they performed SCR because it could improve chemotherapy efficacy [31]. SCR did not have a significant effect on other parameters, such as inflammatory response [30,33].

4. Discussion

Taken together, the results indicate that SCR during chemotherapy is not likely to cause significant adverse effects, and is possible to alleviate treatment-induced side effects, improve quality of life, and stabilize hematological responses. Based on these results, SCRs are worth consideration for larger human trials; however, more high-quality RCTs are necessary before making relevant clinical practice recommendation.

The first important and clear takeaway is that SCR is feasible and well tolerated in cancer patients undergoing chemotherapy, in accordance with researchers who advocate for SCR [8,10]. The side effects directly caused by SCR were rare, and (if any) mild. Though weight loss and malnutrition may be the most worrisome side effects of SCR, the studies show that weight loss is minimal and reversible, and most nutrition parameters (such as prealbumin) remained stable during and after SCR [34]. Despite the minor side effects, the studies' participant retention rates remains a big challenge. In addition, SCR has not yet been thoroughly examined in various types of cancer, male patient groups, and ethnically diverse patient populations. Ethnic or cultural factors play an important role in performing SCR, as eating behavior is closely associated with cultural beliefs [36]. Indeed, some of the reviewed studies showed that one of the barriers to continuing SCR is the social constraint when eating with others, since eating can be considered as a social activity and not only as a means to meet nutritional needs [31,34]. Since only one reviewed study addressed bioelectrical impedance analysis [33], future research may need to consider monitoring nutritional status more aggressively, such as by measuring the change in lean body mass [37].

Corresponding to Lee and Longo's definition of SCR [8,10], the studies that set an SCR regimen required the participants to stay below 50% of the recommended daily calorie intake for no more than a week. Apart from this rough recommendation, it is necessary to discuss whether there is a more precise and appropriate amount and duration of calorie reduction. In the reviewed studies, protocols of reducing calorie intake to zero or providing a 200–400 kcal calorie intake were achievable. However, when compared with a zero-calorie intake, providing a small calorie intake or fasting-mimicking diet caused additional adverse effects, such as aversion or nausea to the provided nutrition [29,38]. On the other hand, the patients showed strong motivation for fasting and indicated that the anxiety of hospitalization automatically lowered their interest in eating [29–31]. Thus, it seems that shortly reducing the calorie intake to nearly zero during chemotherapy can be physically and psychologically acceptable to cancer patients. Future studies are needed to compare the pros and cons of water fasting, low calorie intake (<350 kcal) and fasting-mimicking diet. Although it is outside the scope of this review, comparing the outcomes of different kinds of diet modification is an important future work. For example, it seems that a high-fat, moderate-to-low-protein, and very low-carbohydrate ketogenic diet is effective against cancer [39,40]. Researchers have also proposed that fructose, amino acid, methionine, or serine restriction may have impacts on cancer treatment [4]. With regard to the SCR duration, though all the studies arranged the SCR around chemotherapy, one study that compared 24-, 48-, and 72-h fasting periods showed that groups that fasted for more than 48 h had the least DNA damage in healthy cells [34]. This result is similar to previous findings that show that fasting for longer than 72 h followed by refeeding can protect hematopoietic stem cells from the chemotherapy-induced toxicity and stimulate the proliferation and rejuvenation of old hematopoietic stem cells [41]. More work comparing the effects of different SCR durations are needed.

Our findings show clues regarding one of the aforementioned mechanisms [8,13]—the way SCR selectively protects normal cells from the stress and toxicity of anticancer. Most of the reviewed studies showed that undertaking SCR with chemotherapy, even for as short a period as a few days, could have a protective effect of healthy cells, which results in improving the overall quality of life and alleviating drug-induced side effects, including physical symptoms, nadir, and DNA damage to normal cells [29,30,33–35]. A couple of studies tried to find the association between SCR and tumor cells' sensitivity to anticancer therapy [32–35]. The researchers measured IGF-1 or observed pathological response and imaging reports. Although a decreasing trend in IGF-1 level and a better pathological response were reported [32–35], the researchers did not arrive at a definite conclusion due to the limited number of studies and sample size. Then again, only one of the reviewed studies measured the inflammatory response, and it found no significant change [30]. Thus, it is difficult to conclude whether SCR had the potential to sensitize tumor cells to chemotherapy or facilitate the establishment of an environment against tumor growth. Moreover, while one of the proposed mechanisms of action is related to SCR-induced autophagy, it is imperative to notice the possible two-sided effects of autophagy modulation in tumor cell. Autophagy has the potential to promote tissue regeneration in both normal and tumor cells which may limit the effectiveness of chemotherapy [42]. More studies are needed to (1) explore the mechanism of action, (2) observe biological indicators 48 h or more after fasting, and (3) ensure a sufficient sample size. In addition, using a method that is sensitive to glucose metabolism, such as an FDG-PET/CT scan, may capture the treatment effects more precisely.

A new benefit of SCR has emerged from the results of the qualitative study: SCR improves patients' psychological well-being by empowering them to restore self-control, be proactive, and feel less uncertain and anxious [31]. The positive psychological impacts of fasting have also been observed in healthy women, who experienced an increased sense of achievement, reward, pride, and control [43]. Psychological benefits should be important considerations for future clinical practice and research.

The inherent limitation of this review is the small and narrow study sample. As SCR during chemotherapy is a developing concept, human research has been conducted within the past ten years, and only in certain population (mostly female breast cancer) and geographical areas (U.S.A., Germany, the Netherlands, and France). The generalizability of the results is further precluded because cancer patients with nutritional issues or in a poor condition were automatically excluded from the studies. Because SCR had to be performed with chemotherapy, longer chemotherapy regimens could not be examined.

5. Conclusions

While growing evidence has shown hopeful effects of SCR in in vivo experiments and cancer patients, this study is the first to synthesize current evidence on SCR performance during chemotherapy in humans. Our findings suggest that the harm is manageable and that the benefits are worth investigating. While some RCTs are ongoing [38,44,45], more well-controlled studies with diverse ethnicities and cancer types are needed to confirm the effects of SCR and to refresh nutrition guidelines. A long-term follow-up would provide useful information regarding treatment effects and long-term side effects, yet the researchers need to overcome several challenges, including the low compliance rate. SCR should be an important consideration in the future, as it is cost-effective and potentially linked to many clinical outcomes. For example, SCR may be a solution for managing chemotherapy-related toxicity or hyperglycemia [8,46]. Clinicians' close follow-up on the emerging evidence of SCR would provide perspectives for their current practice.

Author Contributions: Conceptualization, C.-C.T., H.C., T.-C.H., J.-M.L., F.-M.T. and W.-W.W.; methodology, C.-C.T., T.-C.H. and J.-M.L.; formal analysis, C.-C.T. and H.C.; investigation, C.-C.T., H.C. and T.-C.H.; writing—original draft preparation, C.-C.T.; writing—review and editing, H.C., T.-C.H., J.-M.L., F.-M.T. and W.-W.W.; supervision, C.-C.T. All authors have read and agreed to the published version of the manuscript.

Funding: This research was funded by Ministry of Science and Technology of Republic of China (Taiwan), grant number MOST 109-2314-B-002-207.

Conflicts of Interest: The authors declare no conflict of interest.

References

1. De Cabo, R.; Mattson, M.P. Effects of intermittent fasting on health, aging, and disease. *N. Engl. J.* **2019**, *381*, 2541–2551. [CrossRef] [PubMed]
2. Mattson, M.P.; Longo, V.D.; Harvie, M. Impact of intermittent fasting on health and disease processes. *Ageing Res. Rev.* **2017**, *39*, 46–58. [CrossRef] [PubMed]
3. Anton, S.D.; Moehl, K.; Donahoo, W.T.; Marosi, K.; Lee, S.A.; Mainous, A.G.; Leeuwenburgh, C.; Mattson, M.P. Flipping the metabolic switch: Understanding and applying the health benefits of fasting. *Obesity (Silver Spring Md.)* **2018**, *26*, 254–268. [CrossRef] [PubMed]
4. Kanarek, N.; Petrova, B.; Sabatini, D.M. Dietary modifications for enhanced cancer therapy. *Nature* **2020**, *579*, 507–517. [CrossRef] [PubMed]
5. Hammer, M.J.; Voss, J.G. Malglycemia and cancer: Introduction to a conceptual model. *Proc. Oncol. Nurs. Forum.* **2012**, *39*, E275–E287. [CrossRef] [PubMed]
6. Storey, S.; Von Ah, D. Impact of malglycemia on clinical outcomes in hospitalized patients with cancer: A review of the literature. *Proc. Oncol. Nurs. Forum.* **2012**, *39*, 458–465. [CrossRef]
7. Brandhorst, S.; Longo, V.D. Fasting and caloric restriction in cancer prevention and treatment. *Recent Results Cancer Res.* **2016**, *207*, 241–266. [CrossRef]
8. Nencioni, A.; Caffa, I.; Cortellino, S.; Longo, V.D. Fasting and cancer: Molecular mechanisms and clinical application. *Nat. Rev. Cancer* **2018**, *18*, 707–719. [CrossRef]
9. O'Flanagan, C.H.; Smith, L.A.; McDonell, S.B.; Hursting, S.D. When less may be more: Calorie restriction and response to cancer therapy. *BMC Med.* **2017**, *15*, 106. [CrossRef]

10. Lee, C.; Longo, V. Fasting vs dietary restriction in cellular protection and cancer treatment: From model organisms to patients. *Oncogene* **2011**, *30*, 3305–3316. [CrossRef]
11. Icard, P.; Teboul, B.; El Baze, P. A Simple method to optimize the effectiveness of chemotherapy: Modulation of glucose intake during chemotherapy. *Anticancer Res.* **2017**, *37*, 6199–6202. [PubMed]
12. D'Aronzo, M.; Vinciguerra, M.; Mazza, T.; Panebianco, C.; Saracino, C.; Pereira, S.P.; Graziano, P.; Pazienza, V. Fasting cycles potentiate the efficacy of gemcitabine treatment in in vitro and in vivo pancreatic cancer models. *Oncotarget* **2015**, *6*, 18545. [CrossRef] [PubMed]
13. Zhu, Y.; Yan, Y.; Gius, D.R.; Vassilopoulos, A. Metabolic regulation of Sirtuins upon fasting and the implication for cancer. *Curr. Opin. Oncol.* **2013**, *25*, 630. [CrossRef] [PubMed]
14. Lo Re, O.; Panebianco, C.; Porto, S.; Cervi, C.; Rappa, F.; Di Biase, S.; Caraglia, M.; Pazienza, V.; Vinciguerra, M. Fasting inhibits hepatic stellate cells activation and potentiates anti-cancer activity of sorafenib in hepatocellular cancer cells. *J. Cell. Physiol.* **2018**, *233*, 1202–1212. [CrossRef] [PubMed]
15. Panebianco, C.; Adamberg, K.; Adamberg, S.; Saracino, C.; Jaagura, M.; Kolk, K.; Di Chio, A.G.; Graziano, P.; Vilu, R.; Pazienza, V. Engineered resistant-starch (ERS) diet shapes colon microbiota profile in parallel with the retardation of tumor growth in in vitro and in vivo pancreatic cancer models. *Nutrients* **2017**, *9*, 331. [CrossRef]
16. Di Biase, S.; Shim, H.S.; Kim, K.H.; Vinciguerra, M.; Rappa, F.; Wei, M.; Brandhorst, S.; Cappello, F.; Mirzaei, H.; Lee, C. Fasting regulates EGR1 and protects from glucose-and dexamethasone-dependent sensitization to chemotherapy. *PLoS Biol.* **2017**, *15*, e2001951. [CrossRef]
17. De las Peñas, R.; Majem, M.; Perez-Altozano, J.; Virizuela, J.; Diz, P.; Donnay, O.; Hurtado, A.; Jimenez-Fonseca, P.; Ocon, M. SEOM clinical guidelines on nutrition in cancer patients (2018). *Clin. Transl. Oncol.* **2019**, *21*, 87–93. [CrossRef]
18. American Institute for Cancer Research; LIVESTRONG Foundation; Savor Health™. *HEAL Well: A Cancer Nutrition Guide*; 2015; p. 26.
19. Arends, J.; Bachmann, P.; Baracos, V.; Barthelemy, N.; Bertz, H.; Bozzetti, F.; Fearon, K.; Hütterer, E.; Isenring, E.; Kaasa, S. ESPEN guidelines on nutrition in cancer patients. *Clin. Nutr.* **2017**, *36*, 11–48. [CrossRef]
20. Maschke, J.; Kruk, U.; Kastrati, K.; Kleeberg, J.; Buchholz, D.; Erickson, N.; Huebner, J. Nutritional care of cancer patients: A survey on patients' needs and medical care in reality. *Int. J. Clin. Oncol.* **2017**, *22*, 200–206. [CrossRef]
21. Keinki, C.; Seilacher, E.; Ebel, M.; Ruetters, D.; Kessler, I.; Stellamanns, J.; Rudolph, I.; Huebner, J. Information needs of cancer patients and perception of impact of the disease, of self-efficacy, and locus of control. *J. Cancer Educ.* **2016**, *31*, 610–616. [CrossRef]
22. Amano, K.; Maeda, I.; Morita, T.; Tatara, R.; Katayama, H.; Uno, T.; Takagi, I. Need for nutritional support, eating-related distress and experience of terminally ill patients with cancer: A survey in an inpatient hospice. *BMJ Supportive Palliat. Care* **2016**, *6*, 373–376. [CrossRef] [PubMed]
23. Whittemore, R.; Knafl, K. The integrative review: Updated methodology. *J. Adv. Nurs.* **2005**, *52*, 546–553. [CrossRef] [PubMed]
24. Wright, J.G.; Swiontkowski, M.F.; Heckman, J.D. Introducing levels of evidence to the journal. *JBJS* **2003**, *85*, 1–3. [CrossRef]
25. National Heart Lung and Blood Institute; National Institutes of Health; US Department of Health and Human Services. Study Quality Assessment Tools. Available online: https://www.nhlbi.nih.gov/health-topics/study-quality-assessment-tools (accessed on 14 August 2020).
26. Cesario, S.; Morin, K.; Santa-Donato, A. Evaluating the level of evidence of qualitative research. *J. Obstet. Gynecol. Neonatal Nurs. JOGNN/NAACOG* **2002**, *31*, 708–714. [CrossRef]
27. National Heart Lung and Blood Institute; National Institutes of Health; US Department of Health and Human Services. Quality Assessment of Controlled Intervention Studies. Available online: https://www.nhlbi.nih.gov/health-topics/study-quality-assessment-tools (accessed on 14 August 2020).
28. National Heart Lung and Blood Institute; National Institutes of Health; US Department of Health and Human Services. Quality Assessment Tool for Case Series Studies. Available online: https://www.nhlbi.nih.gov/health-topics/study-quality-assessment-tools (accessed on 14 August 2020).

29. Bauersfeld, S.P.; Kessler, C.S.; Wischnewsky, M.; Jaensch, A.; Steckhan, N.; Stange, R.; Kunz, B.; Brückner, B.; Sehouli, J.; Michalsen, A. The effects of short-term fasting on quality of life and tolerance to chemotherapy in patients with breast and ovarian cancer: A randomized cross-over pilot study. *BMC Cancer* **2018**, *18*, 476. [CrossRef]
30. De Groot, S.; Vreeswijk, M.P.; Welters, M.J.; Gravesteijn, G.; Boei, J.J.; Jochems, A.; Houtsma, D.; Putter, H.; van der Hoeven, J.J.; Nortier, J.W. The effects of short-term fasting on tolerance to (neo) adjuvant chemotherapy in HER2-negative breast cancer patients: A randomized pilot study. *BMC Cancer* **2015**, *15*, 652. [CrossRef]
31. Mas, S.; Le Bonniec, A.; Cousson-Gélie, F. Why do women fast during breast cancer chemotherapy? A qualitative study of the patient experience. *Br. J. Health Psychol.* **2019**, *24*, 381–395. [CrossRef]
32. De Groot, S.; Lugtenberg, R.T.; Cohen, D.; Welters, M.J.; Ehsan, I.; Vreeswijk, M.P.; Smit, V.T.; de Graaf, H.; Heijns, J.B.; Portielje, J.E. Fasting mimicking diet as an adjunct to neoadjuvant chemotherapy for breast cancer in the multicentre randomized phase 2 DIRECT trial. *Nat. Commun.* **2020**, *11*, 1–9. [CrossRef]
33. Zorn, S.; Ehret, J.; Schäuble, R.; Rautenberg, B.; Ihorst, G.; Bertz, H.; Urbain, P.; Raynor, A. Impact of modified short-term fasting and its combination with a fasting supportive diet during chemotherapy on the incidence and severity of chemotherapy-induced toxicities in cancer patients-a controlled cross-over pilot study. *BMC Cancer* **2020**, *20*, 1–14. [CrossRef]
34. Dorff, T.B.; Groshen, S.; Garcia, A.; Shah, M.; Tsao-Wei, D.; Pham, H.; Cheng, C.-W.; Brandhorst, S.; Cohen, P.; Wei, M. Safety and feasibility of fasting in combination with platinum-based chemotherapy. *BMC Cancer* **2016**, *16*, 360. [CrossRef]
35. Safdie, F.M.; Dorff, T.; Quinn, D.; Fontana, L.; Wei, M.; Lee, C.; Cohen, P.; Longo, V.D. Fasting and cancer treatment in humans: A case series report. *Aging (Albany NY)* **2009**, *1*, 988. [CrossRef] [PubMed]
36. Levine, C.S.; Miyamoto, Y.; Markus, H.R.; Rigotti, A.; Boylan, J.M.; Park, J.; Kitayama, S.; Karasawa, M.; Kawakami, N.; Coe, C.L. Culture and healthy eating: The role of independence and interdependence in the United States and Japan. *Pers. Soc. Psychol. Bull.* **2016**, *42*, 1335–1348. [CrossRef]
37. Bruggeman, A.R.; Kamal, A.H.; LeBlanc, T.W.; Ma, J.D.; Baracos, V.E.; Roeland, E.J. Cancer cachexia: Beyond weight loss. *J. Oncol. Pract.* **2016**, *12*, 1163–1171. [CrossRef]
38. De Groot, S.; Lugtenberg, R.; Welters, M.; Ehsan, I.; Vreeswijk, M.; Smit, V.; de Graaf, H.; Heijns, J.; Portielje, J.; van de Wouw, A. *Abstract P1-15-20: DIetary REstriction as an Adjunct to Neoadjuvant ChemoTherapy for HER2-Negative Breast Cancer: Final Results from the DIRECT Trial (BOOG 2013-04)*; AACR: Philadelphia, PA, USA, 2019.
39. Oliveira, C.L.; Mattingly, S.; Schirrmacher, R.; Sawyer, M.B.; Fine, E.J.; Prado, C.M. A nutritional perspective of ketogenic diet in cancer: A narrative review. *J. Acad. Nutr. Diet.* **2018**, *118*, 668–688. [CrossRef] [PubMed]
40. Lv, M.; Zhu, X.; Wang, H.; Wang, F.; Guan, W. Roles of caloric restriction, ketogenic diet and intermittent fasting during initiation, progression and metastasis of cancer in animal models: A systematic review and meta-analysis. *PLoS ONE* **2014**, *9*, e115147. [CrossRef] [PubMed]
41. Mendelsohn, A.R.; Larrick, J.W. Prolonged fasting/refeeding promotes hematopoietic stem cell regeneration and rejuvenation. *Rejuvenation Res.* **2014**, *17*, 385–389. [CrossRef] [PubMed]
42. White, E.; DiPaola, R.S. The double-edged sword of autophagy modulation in cancer. *Clin Cancer Res.* **2009**, *15*, 5308–5316. [CrossRef] [PubMed]
43. Watkins, E.; Serpell, L. The psychological effects of short-term fasting in healthy women. *Front Nutr.* **2016**, *3*, 27. [CrossRef] [PubMed]
44. Dorff, T.B.; Shelechi, M.; Kang, I.; Morgan, T.E.; Groshen, S.G.; Yennu, S.; Garcia, A.A.; Quinn, D.I.; Longo, V. *A Randomized Phase II Clinical Trial of a Fasting-Mimic Diet Prior to Chemotherapy to Evaluate the Impact on Toxicity and Efficacy*; American Society of Clinical Oncology: Alexandria, VA, USA, 2018.
45. ClinicalTrials.gov. Available online: https://clinicaltrials.gov/ct2/results?term=chemotherapy%2C+fasting&cond=Cancer&Search=Apply&recrs=b&recrs=a&recrs=f&recrs=d&recrs=e&recrs=m&age_v=&age=1&gndr=&type=&rslt= (accessed on 14 August 2020).
46. Gutiérrez-Salmeán, G.; Ceballos, G.; Meaney, E. Anthracyclines and cardiotoxicity. *Int. J. Cancer Res. Prev.* **2015**, *8*, 515.

nutrients

MDPI

Article

Brief Hospital Supervision of Exercise and Diet During Adjuvant Breast Cancer Therapy Is Not Enough to Relieve Fatigue: A Multicenter Randomized Controlled Trial

William Jacot [1,2,*], Antoine Arnaud [3], Marta Jarlier [1], Claudia Lefeuvre-Plesse [4], Philippe Dalivoust [5], Pierre Senesse [1], Ahmed Azzedine [6], Olivier Tredan [7], Sophie Sadot-Lebouvier [8], Sébastien Mas [1,2], Marion Carayol [9], Jean-Pierre Bleuse [1], Sophie Gourgou [1], Chloé Janiszewski [1], Silene Launay [1], Véronique D'Hondt [1], Géraldine Lauridant [10], Julien Grenier [3], Gilles Romieu [1], Gregory Ninot [1,2] and Laurence Vanlemmens [10]

1 Val d'Aurelle Montpellier Cancer Institute (ICM), 208 Avenue des Apothicaires, Parc Euromédecine, CEDEX 5, 34298 Montpellier, France; marta.jarlier@icm.unicancer.fr (M.J.); pierre.senesse@icm.unicancer.fr (P.S.); sebastien.mas@protonmail.com (S.M.); jean-pierre.bleuse@icm.unicancer.fr (J.-P.B.); sophie.gourgou@icm.unicancer.fr (S.G.); chloe.janiszewski@icm.unicancer.fr (C.J.); silene.launay@icm.unicancer.fr (S.L.); veronique.dhondt@icm.unicancer.fr (V.D.); gilles.romieu@icm.unicancer.fr (G.R.); gregory.ninot@umontpellier.fr (G.N.)

2 Faculty of Medicine, University of Montpellier, Rue du Pr. Henri Serre, 34000 Montpellier, France

3 Sainte-Catherine Institute, 1750 Chemin Lavarin, 84000 Avignon, France; a.arnaud@isc84.org (A.A.); j.grenier@isc84.org (J.G.)

4 Eugène Marquis Center, Rue de la Bataille Flandres-Dunkerque, CS 44229, 35042 Rennes, France; c.lefeuvre@rennes.unicancer.fr

5 Ambroise Paré Hospital, 1 Rue de l'Eylau, 13006 Marseille, France; philippedalivoust@gmail.com

6 Montélimar Hospital, Quartier Beausseret, BP 249-26, 26216 Montélimar, France; ahmed.azzedine@gh-portesdeprovence.fr

7 Léon Bérard Center, 28 Rue Laennec, 69008 Lyon, France; olivier.tredan@lyon.unicancer.fr

8 René Gauducheau Center, Boulevard Jacques Monod, 44805 Saint-Herblain, France; sophie.sadot-lebouvier@ico.unicancer.fr

9 IAPS Laboratory "Impact of Physical Activity on Health", University of Toulon, Avenue de l'Université, 83957 La Garde, France; marion.carayol@univ-tln.fr

10 Oscar Lambret Center, 3 Rue Frédéric Combemale, 59000 Lille, France; g-lauridant@o-lambret.fr (G.L.); l-vanlemmens@o-lambret.fr (L.V.)

* Correspondence: william.jacot@icm.unicancer.fr

Received: 9 September 2020; Accepted: 4 October 2020; Published: 9 October 2020

Abstract: Supervised exercise dietary programs are recommended to relieve cancer-related fatigue and weight increase induced by adjuvant treatment of early breast cancer (EBC). As this recommendation lacks a high level of evidence, we designed a multicenter randomized trial to evaluate the impact of an Adapted Physical Activity Diet (APAD) education program on fatigue. We randomized 360 women with EBC who were receiving adjuvant chemotherapy and radiotherapy to APAD or usual care at eight French cancer institutions. Data were collected at baseline, end of chemotherapy, end of radiotherapy, and 6 months post-treatment. The primary endpoint was the general cancer-related fatigue score using the MFI-20 questionnaire. Fatigue correlated with the level of precariousness, but we found no significant difference between the two groups in terms of general fatigue ($p = 0.274$). The APAD arm has a smaller proportion of patients with confirmed depression at the end of follow-up ($p = 0.052$). A transient modification in physical activity levels and dietary intake was reported in the experimental arm. However, a mixed hospital- and home-based APAD education program is not enough to improve fatigue caused by adjuvant treatment of EBC. Cancer care centers should consider integrating more proactive diet–exercise supportive care in this population, focusing on precarious patients.

Keywords: breast cancer; exercise; diet; education; fatigue; weight; quality of life

1. Introduction

Cancer-related fatigue is the most distressing and common symptom reported by patients undergoing adjuvant therapy for breast cancer (BC) [1–4]. Fatigue provokes an increase in sedentary behaviors, modification of dietary intake, metabolic changes, fat mass increase, depression, and anxiety [5], and can alter cancer prognosis and treatment [6–8]. Exercise and nutrition programs are recommended by experts and medical societies to relieve cancer-related fatigue during active treatment and to prevent an increase in weight [9,10]. Exercise for cancer patients must include both moderate-intensity aerobic exercise and muscle-strengthening exercises [9], and be regular, frequent (at least 2 h per week), and progressive. Nutrition must aim to maintain a healthy weight, promote eating more plant-based foods, limit red and processed meat, energy-dense foods, salt, sugary drinks, and alcohol, and not rely on dietary supplements [11]. This is a particularly relevant effect in the clinical context of BC, as body mass index (BMI) before and after BC diagnosis, and weight gain after diagnosis were associated with increased mortality in recent meta-analyses [7,8]. Nutritional consultations help manage nutritional disorders that worsen fatigue, such as anemia, diarrhea, nausea, and vomiting [10,12].

The combination of exercise and dietary support has been reported to induce significant weight loss in survivors of early BC (EBC) after adjuvant chemotherapy/radiotherapy [13–15]. Four randomized controlled trials (RCTs) have assessed the combination of exercise and diet in EBC patients undergoing adjuvant chemotherapy and/or radiotherapy, but they were designed as pilot trials with less than 30 patients in each randomization group [16–19]. Therefore, the benefits of an exercise-diet intervention during adjuvant chemotherapy and radiotherapy need to be evaluated in a well-powered and multicenter RCT. We previously reported a monocentric RCT evaluating an adapted physical activity and diet (APAD) intervention during adjuvant treatment of EBC [18,20] and found a beneficial effect on patient-reported outcomes (PROs). However, these results need to be validated in larger, multicenter cohorts in order to evaluate the impact of the heterogeneity of organizations and sociocultural parameters on the results of the intervention.

As social deprivation has been reported to significantly impact cancer risk and program efficacy, this variable needs to be addressed in order to identify the impact of social inequalities on the performance of a given intervention. For example, vulnerable individuals identified using the validated Evaluation of Deprivation and Health Inequalities in Public Health Centers (EPICES) index are more likely already to have cancer, a higher mean BMI, greater prevalence of current smoking, lower adherence to screening programs, and greater standardized mortality ratio compared to non-vulnerable individuals [21–23]. Thus, this social grading variable is expected to impact the adherence of patients to supportive care programs and ultimately induce differences in treatment-induced fatigue in this population.

We designed the present multicenter RCT to assess the effect of a combined exercise and diet intervention delivered during a six-cycle adjuvant chemotherapy regimen followed by radiotherapy on cancer-related fatigue in EBC patients. We hypothesized that the combined supervised program would yield beneficial effects compared to usual care on cancer-related fatigue as a primary outcome, especially in patients considered to be more vulnerable in terms of social deprivation. Secondary outcomes include BMI, nutritional parameters, physical abilities, anxiety-depressive symptoms, and health-related quality of life.

2. Materials and Methods

2.1. Design

The present study was a two-arm, multicenter, randomized, controlled, prospective trial. The APAD2 trial was designed and implemented to evaluate the impact of an exercise and nutrition-based supportive care intervention during 6 months of chemotherapy and radiotherapy on fatigue, evaluated using the MFI-20 questionnaire in EBC patients treated in eight French cancer centers. The APAD intervention was compared to usual care without specific exercise and/or nutrition care (control arm). The primary hypothesis was the possibility of obtaining a 4-point reduction in the mean score on the General fatigue subscale in the intervention group with respect to the control group.

2.2. Participants

Eligible participants were women over 18 years of age with histologically proven and newly (<6 months) affected by EBC accessible to initial surgery, without consideration of their baseline physical activity level or dietary intake. Patients were enrolled after undergoing curative surgery. All patients were to receive six cycles of adjuvant chemotherapy (three cycles of epirubicin/cyclophosphamide/5-fluorouracil every 3 weeks (FEC100 protocol), followed by three taxane-based cycles, either docetaxel every 3 weeks [24] or paclitaxel weekly for 9 weeks), followed by 6 weeks of radiotherapy. Patients affected by HER2-positive tumors also received adjuvant trastuzumab for a total of 52 weeks, starting at the initiation of taxane chemotherapy. Exclusion criteria were metastatic disease, any other primary tumor, medical contra-indications to moderate-intensity physical activity, inability to attend intervention sessions or assessments, and a difficulty or disability preventing the patient from correctly understanding the trial information or requirements. The study was approved by the local ethics committee (NCT04109326) and conducted in accordance with the Declaration of Helsinki and principles of good clinical practice.

Before chemotherapy, potential participants were identified by the hospital medical oncologists. All participants were informed of the goal of the study and the potential benefits of diet and exercise on fatigue during adjuvant therapy. The patients who provided written informed consent and completed baseline assessments were randomly assigned (1:1 ratio) to the APAD experimental arm or control arm, stratified by center and precariousness level as assessed by the EPICES score [22] using the minimization method. Randomization was performed at the Montpellier ICM Biometric Unit using a computer program generated with Stata software version 12 (StatCorp, LLC, College Station, TX, USA). Participants, interventionists, and assessors were not masked to group assignment. The control group was a usual care group without any diet or exercise intervention. No particular material was delivered to the control group during the intervention, and these patients were not asked to limit exercise practice or eat/avoid specific foods during the intervention period.

2.3. Intervention

The APAD education program, which was based on the previously published trial method [18], was implemented during chemotherapy and radiotherapy (26 weeks). The intervention included twice-weekly exercise sessions and six individual nutritional therapeutic education sessions. The exercise sessions were individually supervised hospital-based exercise sessions and non-supervised home-based sessions combining one muscle strength session and one aerobic session each week (Figure 1). The nutritional education sessions targeted body weight control and the modification of feeding behaviors according to the WCRF recommendations [25]. The nutritional sessions were planned on the same days as supervised hospital-based exercise sessions. The intervention was tailored to maintain the patients' exercise level and dietary intake in accordance with the guidelines throughout the intervention period.

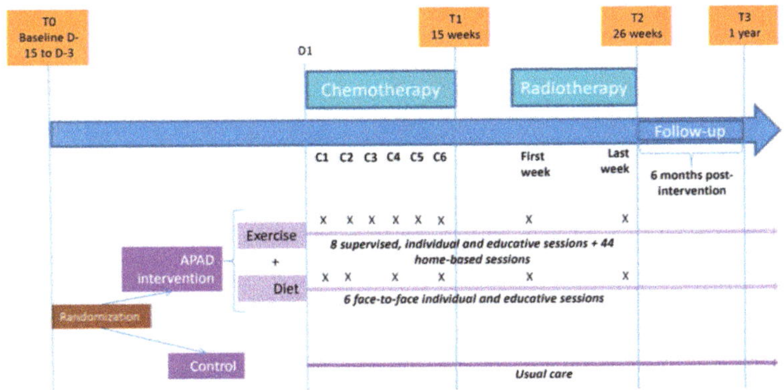

Figure 1. Flow diagram of the Adapted Physical Activity Diet 2 (APAD2) trial. C indicates chemotherapy cycle time points. X indicates intervention times in the associated randomization arm. Number of participating patients: T0 (Baseline): Control (n = 180), Adapted Physical Activity Diet (APAD) (n = 180). T2 (completed Chemotherapy and Radiotherapy sessions): Control (n = 170), APAD (n = 160). T3 (1 year after inclusion): Control (n = 157), APAD (n = 144).

2.3.1. Exercise

The exercise program delivered by trained professionals combined aerobic and muscle strengthening exercises according to international recommendations: 120 min of moderate to vigorous physical activity per week associated with strength training [9,26–28]. One muscle strength session and one aerobic session were scheduled each week (10 min of warm-up, at least 30 min of exercise, 10 min of stretching and 10 min of relaxation time with a goal of muscle recovery and well-being). Strength sessions targeted six main muscle groups (hamstrings, quadriceps, buttocks, abdominal, back, shoulders/arms), and each skill was performed for 2 to 5 sets with 6 to 12 repetitions with individual adaptation and progression. Aerobic exercise was performed at moderate intensity and adapted to the patient's physical condition and progression, with a target of 50–75% of the maximum heart rate for 30 to 45 min (adapted to the patient's physical condition). The initial exercise intensity was individualized but generally began at 50–55% of the maximum heart rate and progressed to 65–75% of the maximum heart rate by weeks 20 to 26. Supervised hospital-based sessions were achieved on a cycloergometer. For home-based practice, patients were given the option of various modalities of aerobic exercise (e.g., walking, jogging, cycling, dancing/fitness, swimming) to sustain adherence to the program and promote enjoyment. Hospital-based supervised exercise sessions aimed to provide the patients with relevant instructions that allow reproducibility at home and increased autonomy. Every supervised session was based on theory-based behavioral targets and techniques to improve behavioral change and patient adherence. Hospital-based supervised exercise sessions were scheduled on the same day as chemotherapy and during radiotherapy every 3 weeks to avoid additional cost. A total of eight hospital-based supervised exercise sessions and 44 at-home sessions were planned during the course of the intervention. Non-supervised, home-based sessions were planned at least twice per week, except only one home-based session was scheduled for the weeks that included one supervised hospital-based exercise session. Precise written instructions were given to patients in the educational and personable APAD-Moving workbook, which included information on their disease and reasons for being physically active during active treatment for cancer, illustrated instructions on performing the home exercises, the schedule for planned home-based sessions, and a patient log to evaluate adherence. Patients were asked to fill in the adherence log at home with whether planned sessions were achieved, the number of achieved muscular exercises, duration of each session, rating of perceived exertion (on scale of 1 to 10), reason for missed sessions, and anything else that they would like to discuss with the exercise specialist at the next supervised session.

2.3.2. Diet

Patients in the intervention arm received diet counselling with therapeutic education from a dietician at six individual face-to-face sessions. Each session lasted approximately 30 min. During chemotherapy, four diet sessions were scheduled to achieve balanced dietary intake, advising patients on controlling weight and managing with the potential toxicities and side effects of chemotherapy. Two more sessions were planned at the beginning and end of radiotherapy for all intervention groups. Weight control was pursued in patients with BMI < 30 kg/m^2; weight normalization was targeted in patients with BMI ≥ 30 kg/m^2 (i.e., to decrease BMI to less than 30 kg/m^2 by the end of adjuvant therapy). Each consultation involved an evaluation of nutritional status, nutrition care tailored to the patient's caloric needs and potential toxicities related to treatment, and nutritional education. The purpose of these consultations was to teach the principles of a well-balanced diet, foster weight control during treatment, and induce appropriate feeding behaviors after treatment.

2.3.3. Evaluation of Nutritional Status

At the first session, nutritional status was evaluated based on the patient's usual weight, current weight, and their weight measured 1 to 6 months prior to study enrollment according to the French National Authority for Health criteria. The dietician assessed the patient's daily energy requirement by computing their basal metabolic rate (BMR) [29] according to the corrected formula of Harris and Benedict. Dietary intake was prospectively measured by asking patients to fill out a questionnaire on 3 consecutive days of food intake at the first and last session. For the other sessions, dietary intake was measured by a 24-h recall food survey and a 10-point visual analogue scale.

2.3.4. Nutrition Care

Nutrition care aimed for weight control through balanced dietary intake tailored to the patient's energy needs and potential toxicities related to the cancer treatment. The dietician verified the patient's intake utilizing the following guidelines: daily energy intake was compared to the estimated daily energy needs, patients were guided to regularly distribute their dietary intake into three main meals with an optional snack in the afternoon, macronutrient distribution was compared to the French dietary reference intakes for a balanced diet (i.e., 30–35% lipids, 50–55% carbohydrates, and 10–15% protein) [30], and food group intake was guided to meet the recommendations of the WCRF [25]. If the patient's habits did not correspond with these guidelines, or their daily energy intake was higher or lower than 10% of the estimated daily energy needs, the dietician counselled the patient on modifications regarding foods, nutrients, meals, and calorie distribution. If the patient's BMI was >30 kg/m^2 by the end of chemotherapy, a new weight goal was set to decrease the patient's BMI to the range of 25 to 30 kg/m^2. A new range of daily energy needs was then estimated with a corresponding distribution according to food group balance and the WCRF guidelines [25]. Patients were given a printed example of food groups, servings, and distribution that they may eat on a typical day. In the following sessions, the dietician computed the patient's intake again and adapted the advice to the evolution of the patient's intake. Specific advice was given to patients on the management of potential toxicities and side effects of chemotherapy.

2.3.5. Nutritional Education

Nutritional education aimed to teach the patients the principles of a well-balanced and healthy diet based on WCRF guidelines [25], inform them of industrial food packaging, and fight preconceived ideas by using practical applications and educational games. Nutritional education was tailored to the patients' habits and means, precariousness level, and cultural and social environment.

Session 1: Presentation of detailed well-balanced menus.

Session 2: Identify the nature and specific roles of food groups—food balance education based on the food pyramid presentation.

Session 3: Food balance education based on the "APAD fridge game" consisting of the elaboration of three balanced meals with the food provided on the picture (food and dish choices from proposed menus to obtain balanced meals in special contexts, such as picnic, fast food, or restaurant).

Session 4: Teach to read labels and food packaging—food balance education through examples of complex mixed dishes (e.g., lasagna).

Session 5: Evidence-based information on the relationship between nutrition and cancer using a quiz game pointing out preconceived ideas.

Session 6: Post-treatment diet benefits and recommendations, and delivery of a booklet summarizing dietary WCRF recommendations.

2.3.6. Missed Sessions

Missed supervised exercise or diet counseling sessions at the hospital during chemotherapy could not be rescheduled, as the patients only came to the hospital once every 3 weeks during chemotherapy. In the case of a missed supervised session, a phone call was made to the patient by the exercise and/or diet specialists. The discussion focused on reasons for not attending the session, patient adherence in the last 3 weeks, encouraging the patient to attend future exercise or diet counseling sessions, taking into account the difficulties, and delivering education targets and content of the missed session if possible.

In contrast, missed supervised hospital-based sessions during radiotherapy were rescheduled as soon as possible because most of the patients came to the hospital every weekday during radiation therapy.

2.4. *Outcomes and Assessments*

The endpoints and assessments were concordant with those of the previous APAD1 trial [18,20]. Endpoints included subjective PROs and objective outcomes. Assessments were conducted at each site at baseline, just before the start of adjuvant chemotherapy (T0); the end of chemotherapy (T1); end of radiotherapy (T2); and the 6-month follow-up (T3). The primary endpoint was self-reported cancer-related fatigue assessed by the General fatigue subscale of the Multidimensional Fatigue Inventory (MFI-20) [31,32], a 20-item self-report instrument that covers five dimensions. The other four subscales (Physical fatigue, Mental fatigue, Reduced activity, and Reduced motivation) and the total score were considered secondary outcomes.

An objective measure of physical fatigue, lower limb muscle endurance, was measured using the sit-to-stand test at 15 and 30 s. The adherence to chemotherapy was evaluated using the relative dose intensity (RDI), which was calculated as the ratio of the cumulative dose intensity (mg/m^2/week) to the dose intensity planned in the chemotherapy protocol. In addition, anxiety and depression symptomatology was evaluated using the 14-item Hospital Anxiety Depression Scale (HADS) self-report questionnaire [33]. Quality of life (QoL) was assessed by the EORTC QLQ-C30 questionnaire, a validated cancer-specific instrument [34] evaluating five functions (physical, role, cognitive, emotional, and social), nine symptoms (fatigue, pain, nausea and vomiting, dyspnea, loss of appetite, insomnia, constipation, diarrhea, and financial difficulties), and global health status. Physical activity was assessed using the 16-item Global Physical Activity Questionnaire (GPAQ) developed by the World Health Organization (WHO) [35,36]. The GPAQ assesses the intensity, duration, and frequency of physical activity in a usual week in three domains: activity at work, travel to and from places, and recreational activities. Physical activity is then expressed in terms of metabolic equivalent (MET), which is the ratio between the speed of metabolism during physical activity and the speed of metabolism at rest [37]. MET values are applied to vigorous and moderate intensity variables in work and recreational settings. One MET is defined as 1 kcal/kg/h and is equivalent to the energy cost of sitting quietly. We attribute 4 MET and 8 MET to the time spent on moderately intense and vigorous physical activity, respectively. Outcomes considered were average METs (MET/min/day) from activities of moderate and vigorous intensity (work and recreational); average METs from moderate intensity transport (cycling and walking), total physical activity (MET-minutes/week), and sedentary time (min/day). We also calculated the proportion of patients with a low level of total physical activity and the proportion of patients that failed to meet the WHO recommendations. Anthropometric measures (body weight, height, BMI, and waist circumference) were used to describe weight gain. Dietary intake was evaluated using a food record [38] of the foods and beverages consumed for 3 consecutive days (including one weekend day); the data were entered into nutritional analysis software and calories and nutrient intake computed (Nutritional Analysis Software, release 8, Villiers-les-Nancy: MICRO 6, 2007). All endpoints, except the nutritional evaluation, were assessed at four time points: pre-intervention, baseline assessment before the start of adjuvant chemotherapy (T0); end of chemotherapy (T1); end of radiotherapy (T2; i.e., immediately post-intervention); and 6 months after the end of treatment (i.e., 1 year after inclusion in the study). The nutritional evaluation was conducted at three time points: T0, T2, and T3.

2.5. Statistical Considerations

2.5.1. Sample Size Calculation and Randomization

The sample size calculation was based on the primary endpoint, the General fatigue subscale of the MFI-20 questionnaire. Smets et al. [31] estimated a mean score of 16 (standard deviation [SD] 8) for the General fatigue subscale in cancer patients undergoing radiotherapy; therefore, we considered this the reference value and the main time point of the evaluation was T2. In order to detect a 4-point reduction in the mean score on the General fatigue subscale in the intervention group with respect to the control group (i.e., reduction of 25% of the reference), the sample size calculation was based on a global risk alpha of 1% (bilateral situation), 90% (1-β) power, SD of 9, and the consideration of a repeated measures design (one pre-intervention measure and roughly three post-intervention measures, with a hypothetical correlation coefficient between measures of 0.2). The sample size was estimated to be 161 patients per arm in Stata Statistical Software, Release 10 (StataCorp, College Station, TX, USA). Considering a 12% loss to follow-up, 180 patients were required per group, a total of 360 patients overall in the trial. Randomization (1:1) was achieved using the minimization method and patients stratified according to two factors: the level of socioeconomic deprivation assessed by the French EPICES score [21,22] and recruiting center. Randomization was centralized at the Biometrics Units of the ICM, Montpellier.

2.5.2. Statistical Analysis

All analyses were performed in the intention-to-treat population. Analyses related to the impact of the program were also performed in the per-protocol population. This population considers all eligible patients treated and evaluated, which in the case of the intervention arm was defined as all patients who completed at least one supervised or unsupervised exercise session. An initial descriptive analysis of the baseline variables was performed and balance between treatment groups checked for the main demographic, socio-professional, clinical, and PRO variables.

The efficacy of the program was evaluated by the relative difference in the General fatigue score at each time point (T1, T2, and T3) with respect to T0 and compared between two groups using the Kruskal–Wallis test. A model approach was also used; the evolution of the General fatigue score over time was assessed using a linear mixed model (LMM). Random intercepts and random slopes were included to take into account the time effect. Interaction terms were also considered. The model coefficients were estimated by maximum likelihood; coefficients are presented as β_1 for arm effect (APAD with respect to control) and as β_0 for time effect. Separate models were adjusted for each other of the MFI-20 questionnaire subscales and for all other secondary endpoints.

The EORTC QLQ-C30 was analyzed following the EORTC guidelines [39], with scores for each functional and symptom subscale. HADS scores assessing anxiety and depression were categorized according to Zigmond classification (absence of disorder, suspected disorder, disorder).

Physical activity data recorded with the GPAQ were analyzed according to the WHO guidelines and using the STEPS analysis syntax program for cleaning and analyses.

Categorical and ordinal variables were compared using the Pearson chi-squared or Fisher's exact tests. Continuous variables were presented as mean and standard deviation (SD) and were compared using the non-parametric Kruskal–Wallis test.

In the case of missing values, no imputation method was used. All reported p-values are two-sided, and a significance threshold of 0.05 was considered. Statistical analyses were performed using the STATA 13 software (StataCorp LP, College Station, TX, USA).

2.6. Ethics Approval, Consent to Participate and Trial Registration

The study was approved by the French institutional review board (i.e., Comité de Protection des Personnes Sud-Méditerranée III; N°ID-RCB: 2012-A01648-35), the Agence Nationale de Sécurité du Médicament et des produits de santé (N°ANSM: 130313B-12), and the Commission nationale de l'informatique et des libertés. Written informed consent was obtained from all participants. The trial was registered under the identification number NCT04109326.

3. Results

From May 2013 to December 2014, 360 patients were randomized at eight centers to the APAD (n = 180) and control (n = 180) arms. PROs were collected from 99%, 85%, 81%, and 71% of the 360 randomized patients at T0, T1, T2, and T3, respectively (Figure 2). Thirty-eight patients in the APAD arm (21%) did not complete any of the supervised or unsupervised exercise sessions. Among the per-protocol APAD population (n = 142), 67 patients (47.2%) had 80% adherence to the exercise program (defined as the completion of at least 80% of supervised and 80% of unsupervised sessions); 93 patients (65.5%) had global 80% adherence (both modalities confounded).

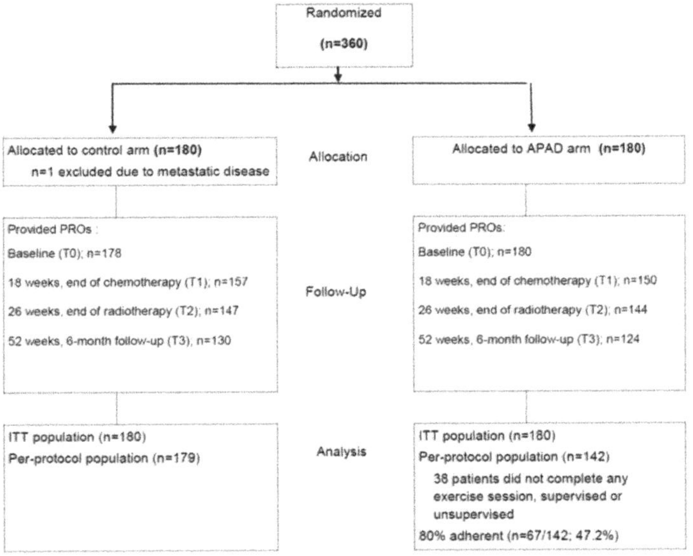

Figure 2. CONSORT diagram of the study.

The baseline characteristics of the patients are presented in Table 1. Most of the patients completed the study (80% in the APAD arm, 87.7% in the control arm). The study was discontinued early for 17% and 7.2% of the patients in the APAD and control arm, respectively, due to patient withdrawal from the study (p = 0.014).

Table 1. Baseline characteristics of patients in the intention-to-treat population.

	Control n = **180**		**APAD** n = **180**		**Total** n = **360**	
	Mean	SD	Mean	SD	Mean	SD
Age (years)	52.35	10.09	52.66	9.69	52.51	9.88
Weight (kg)	67.00	14.13	68.41	14.60	67.71	14.36
BMI (kg/m^2)	25.22	5.30	25.72	5.14	25.47	5.22
BMI categories	n	%	n	%	n	%
<18.5 kg/m^2	6	3.33	2	1.12	8	2.23
18.5–24.9 kg/m^2	99	55.00	95	53.07	194	54.04
25–29.9 kg/m^2	45	25.00	45	25.14	90	25.07
≥30 kg/m^2	30	16.67	37	20.67	67	18.66
Post-menopausal	88	48.89	88	48.89	176	48.89
Tobacco smoking						
Non-smoker	102	56.67	86	47.78	188	52.22
Smoker	29	16.11	34	18.89	63	17.50
Ex-smoker	49	27.22	60	33.33	109	30.28
Marital status						
Single/divorced/widowed, no child	16	8.99	10	5.56	26	7.26
Single/divorced/widowed, with child	23	12.92	37	20.56	60	16.76
Married/living together, no child	21	11.80	26	14.44	47	13.13
Married/living together, with child	118	66.29	107	59.44	225	62.85
Education level						
No qualifications	29	16.57	24	13.56	53	15.06
Secondary level	43	24.57	31	17.51	74	21.02
Completed high school	29	16.57	43	24.29	72	20.45
Completed ≥ 2 years at university	74	42.29	79	44.64	153	43.47
Usual professional status						
Full or part-time employed	97	53.89	103	57.22	200	55.56
Retired	42	23.33	41	22.78	83	23.06
Unemployed/medical leave	41	22.78	36	20.00	77	21.38
EPICES precariousness (or deprivation) level						
Non-precarious	109	60.56	109	60.56	218	60.56
Intermediate	60	33.33	60	33.33	120	33.33
Precarious	11	6.11	11	6.11	22	6.11
Surgery type	n	%	n	%	n	%
Lumpectomy	89	49.44	88	48.89	177	49.17
Quadrantectomy	37	20.56	45	25.00	82	22.78
Mastectomy	54	30.00	46	25.56	100	27.78
T stage						
T1	91	50.56	97	53.89	188	52.22
T2	74	41.11	73	40.56	147	40.83
T3	11	6.11	8	4.44	19	5.28
T3	3	1.67	1	0.56	4	1.11
T4	1	0.56	0	0	1	0.28
Tis	0	0	1	0.56	1	0.28

Table 1. *Cont.*

	Control $n = 180$		APAD $n = 180$		Total $n = 360$	
T stage						
N0	71	39.66	79	44.63	150	42.13
N1	86	48.04	83	46.89	169	47.47
N2	14	7.82	11	6.21	25	7.02
N3	7	3.91	3	1.69	10	2.81
NX	1	0.56	1	0.56	2	0.56
Breast cancer subtype						
Triple negative	17	18.48	17	18.89	34	18.68
HER2+, ER+, and/or PR+	29	31.52	35	38.89	64	35.16
HER2+, ER−, and PR−	9	9.78	10	11.11	19	10.44
HER2−, ER+, and/or PR+	37	40.22	28	31.11	65	35.71

3.1. Fatigue

Compared to T0, the median relative difference in General fatigue scores at T1 was 25% in the control arm and 21% in the APAD arm, which is not a significant difference between the two arms ($p = 0.274$; Table 2). At T2, the increase in fatigue was greater in the APAD arm (20%) than in the control arm (8%), but the difference was still not significant ($p = 0.157$). However, at T3, 1 year after inclusion, the increase in fatigue in the APAD arm (15%) was lower than that of the control arm (20%), though it was not statistically or clinically significant ($p = 0.933$, Figure 3). According to the adjusted model, general fatigue tends to increase over time ($\beta_0 = 0.024$ [95% CI 0.01; 0.034]; $p < 005$) without an observable effect of the intervention in terms of a reduction in general fatigue over time ($\beta_1 = 0.33$, $p = 0.374$; Table 2).

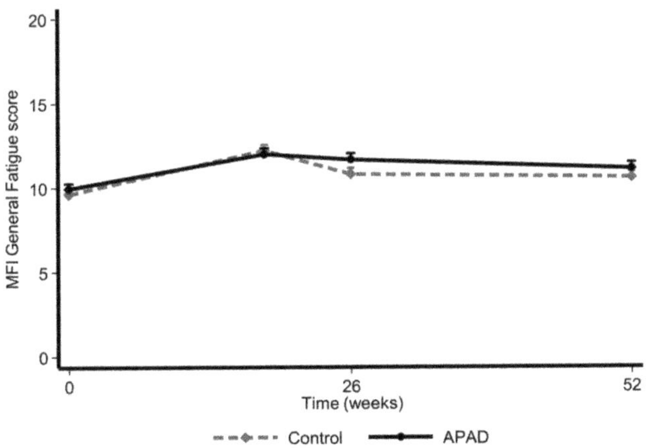

Figure 3. Evolution of the MFI General Fatigue score according to randomization arm in the intention-to-treat population. Data are presented as mean + SD.

Table 2. Fatigue sub-scales of the MFI20 and quality of life (EORTC QLQ-C30) in the intention-to-treat population.

		Baseline (T0)		End of CT (T1)			End of RT (T2)			1 Year after Inclusion (T3)			LMM Coefficients [1] (95% CI)
		Mean	SD	Mean	SD	p	Mean	SD	p	Mean	SD	p	
General fatigue (Endpoint)	Control	9.63	4.2	12.19	4.21	0.683	10.79	4.44	0.107	10.58	3.84	0.231	$\beta_1 = 0.33$ [−0.40; 1.07], $p = 0.374$
	APAD	9.95	4.18	11.97	4.37		11.65	4.44		11.1	4.07		$\beta_0 = 0.24$ [0.014; 0.034], $p < 0.05$
Median (range) of the relative difference[2]	Control			0.25 (−0.67; 3.0)			0.083 (−0.60; 3.50)			0.20 (−0.69; 3.5)			
	APAD			0.21 (−0.5; 2.75)			0.20 (−0.58; 3.25)			0.15 (−0.6; 3.0)			
				$p = 0.274$			$p = 0.157$			$p = 0.933$			
Physical fatigue	Control	9.55	3.72	11.28	4.05	0.896	10.61	3.96	0.985	10.1	3.63	0.742	$\beta_1 = -0.15$ [−0.81; 0.52], $p = 0.670$
	APAD	9.11	3.85	11.33	4.19		10.64	4.25		10.06	3.77		$\beta_0 = 0.02$ [0.007; 0.025], $p < 0.05$
Mental fatigue [3]	Control	7.66	3.81	9.19	4.26	1	8.9	4.31	1	8.82	4.26	0.872	$\beta_2 = -0.015$ [−0.035; 0.005], $p = 0.152$
	APAD	8.57	4.07	9.15	4.24		9.06	4.71		9.06	4.65		$\beta_0 = 0.03$ [0.013; 0.042], $p < 0.05$
Reduced activities	Control	8.34	3.64	10.12	4.26	0.466	9.14	3.92	0.433	8.75	4.02	0.743	$\beta_1 = 0.30$ [−0.40; 0.99], $p = 0.399$
	APAD	9.01	4.01	9.84	4.64		9.67	4.49		8.9	4		$\beta_0 = 0.004$ [−0.005; 0.014], $p = 0.402$
Reduced motivation [3]	Control	7.73	3.5	8.22	3.92	0.587	7.84	3.46	0.651	8.07	3.19	0.572	$\beta_1 = -0.30$ [−0.34; 0.93], $p = 0.360$
	APAD	8.54	3.73	8.31	3.7		8.15	3.77		8.1	3.83		$\beta_0 = 0.002$ [−0.006; 0.009], $p = 0.696$
EORTC QLQ-C30													
Global health status	Control	69.94	18.67	59.39	21.22	0.516	66.55	19.53	0.429	67.44	19.3	0.537	$\beta_1 = -0.0026$ [−0.071; 0.065], $p = 0.940$
	APAD	68.45	19.55	60.83	21.34		64.82	19.1		69.17	17.76		$\beta_0 = 0.001$ [−0.001; 0.001], $p = 0.824$
Physical functioning	Control	87.17	14.59	79.79	19.1	0.219	84.4	15.38	0.083	85.45	17.26	0.194	$\beta_1 = -0.020$ [−0.034; 0.075], $p = 0.466$
	APAD	88.48	14.64	82	19.58		86.49	17.11		89.46	12.95		$\beta_0 = -0.0001$ [−0.0007; 0.0006], $p = 0.873$
Role functioning	Control	84.46	22.16	77.99	23.34	0.257	84.35	21.03	0.966	87.05	18.48	0.309	$\beta_1 = 0.026$ [−0.073; 0.125], $p = 0.607$
	APAD	86.57	20.4	80.63	23.25		83.8	22.01		90.05	16.86		$\beta_0 = 0.0009$ [−0.0005; 0.0022], $p = 0.220$
Emotional functioning	Control	63.7	23.52	72.75	24.83	0.211	75.51	23.09	0.309	73.26	20.1	0.679	$\beta_1 = 0.007$ [−0.106; 0.119], $p = 0.910$
	APAD	63.66	23.95	70.5	22.93		73.1	23.28		73.21	23.33		$\beta_0 = 0.004$ [0.002; 0.005], $p < 0.01$
Cognitive functioning	Control	85.21	20.5	79.3	25.14	0.657	80.16	21.75	0.849	79.97	23.42	0.957	$\beta_1 = -0.019$ [−0.114; 0.076], $p = 0.695$
	APAD	84.17	20.32	78.56	24.37		79.58	22.4		80.24	23.12		$\beta_0 = -0.002$ [−0.004; −0.001], $p < 0.01$
Social functioning	Control	82.58	24.15	66.77	30.63	0.827	73.58	27.71	0.23	82.04	23.99	0.849	$\beta_1 = 0.003$ [−0.128; 0.134], $p = 0.964$
	APAD	84.45	21.97	66.44	29.73		69.84	28.42		83.6	21.78		$\beta_0 = -0.00008$ [−0.002; 0.002], $p = 0.942$
Fatigue symptom	Control	28.54	21.86	44.87	27.24	0.934	34.62	25.73	0.152	32.26	21.96	0.935	$\beta_1 = -0.042$ [−0.248; 0.164], $p = 0.687$
	APAD	28.49	23.35	45.38	29.13		37.4	23.8		31.63	21.15		$\beta_0 = 0.004$ [0.001; 0.007], $p = 0.018$

[1] In the linear mixed model (LMM; log transformed variables): β_1 is the coefficient of the variable 'arm' (interpreted as APAD effect with respect to Control) noted as β_2 when interaction (arm by time) was significant. β_0 is the coefficient of the variable time (weeks) variable. [2] For each patient, the relative difference (RD) with respect to the baseline value of the General fatigue subscale at the end of chemotherapy (CT), end of radiotherapy (RT; end of the oncological treatment), and 1 year after inclusion was calculated as: ($GFS_{\_endRT}$ – $GFS_{\_Inclusion}$)/$GFS_{\_Inclusion}$. A smaller RD indicates a greater reduction in general fatigue. [3] A baseline imbalance between arms was observed for Mental fatigue ($p = 0.019$) and Reduced motivations ($p = 0.025$). No other baseline imbalance was observed.

No significant benefit of the APAD program was observed when the analysis was conducted by precariousness level (EPICES score). However, we observed a clear gradient in the baseline fatigue according to the stratum of precariousness, with the fatigue level being higher for the most precarious stratum throughout the study (Figures S1 and S2).

For all other fatigue subs-scales of the MFI20, no difference was observed between treatment arms with respect to their evolution over time. Overall, physical and mental fatigue tended to increase across time in both arms ($\beta_0 = 0.02$ and $\beta_0 = 0.03$, $p < 0.05$; Table 2). Notably, a baseline imbalance disfavoring the APAD arm was observed for Mental fatigue ($p = 0.019$) and Reduced motivations ($p = 0.025$).

When the fatigue sub-scales were analyzed in the per-protocol population (Table S1), we observed that the APAD program had a positive impact, enhancing motivation over time despite the initial imbalance ($\beta2 = -0.023$ [95% CI −0.39; −0.008]; $p = 0.003$), and reducing mental fatigue ($\beta2 = -0.020$ [−0.04; 0.002], $p = 0.069$).

In a sub-group analysis considering the 80% adherent population in the APAD and control arms, we found no significant difference in the relative differences in General fatigue scores (primary endpoint) (data not shown).

3.2. *QoL and Psychological Distress*

Quality of life and psychological distress scores on the EORTC QLQ-C30 and HADS, respectively, at baseline and during follow-up are described in Tables 2 and 3. At the end of radiotherapy, we found no significant difference between arms in terms of QoL (Table 2, Figure S3). Overall, emotional function (both arms together) increased (LMM coefficient $\beta_0 = 0.004$ [0.002; 0.005], $p < 0.01$) and cognitive function decreased ($\beta_0 = -0.002$ [−0.004; −0.001], $p < 0.01$) over time. The symptom fatigue also increased ($\beta_0 = 0.004$ [0.001; 0.007], $p = 0.018$), which is consistent with the results observed in the MFI score analysis. No impact of the APAD program was observed in regard to improving QoL dimensions global health status, functional dimensions, and fatigue symptoms.

When the analysis was conducted on the per-protocol population (Table S1 and Figure S4), we observed slightly better physical function at the end of radiotherapy ($p = 0.043$), but the LMM did not show a global trend over time for an impact of the APAD program on this function ($\beta1 = 0.025$ [−0.034; 0.084]; $p = 0.800$).

In a sub-group analysis considering the 80% adherent population in the APAD and control arms, we observed significantly better physical function in the APAD subgroup at T2 ($p = 0.003$) and T3 ($p = 0.017$). However, in that subgroup analysis, a baseline imbalance favored the 80% adherent APAD subgroup, which exhibited better physical function ($p = 0.028$) and lesser pain symptom ($p = 0.002$) at T0. This may be explained, in part, to better adherence to the program.

Regarding psychological distress (Table 3), a lower proportion of patients with confirmed depression (score >10) was observed in the APAD arm (54.03%) with respect to control (66.92%) at T3 ($p = 0.052$). A similar result was observed in a per-protocol analysis ($p = 0.026$; Table S2).

Table 3. Anxiety and depression disorders in the intention-to-treat population.

	Control		APAD		
	n	%	*n*	%	*p*
Baseline (T0)					
Anxiety					0.662
Absence (<7)	0	0.00	0	0.00	
Suspected (8–10)	2	1.12	3	1.67	
Confirmed (>10)	176	98.88	177	98.33	
Mean anxiety (SD)	11.85 (2.56)		11.92(2.69)		
Depression					0.368
Absence (<7)	0	0.00	2	1.11	
Suspected (8–10)	70	39.33	71	39.44	
Confirmed (>10)	108	60.67	107	59.44	
Mean depression (SD)	18.93 (3.34)		18.78 (3.54)		
End of chemotherapy (T1)					
Anxiety					0.974
Absence (<7)	0	0.00	0	0.00	
Suspected (8–10)	1	0.64	1	0.67	
Confirmed (>10)	156	99.36	149	99.33	
Mean anxiety (SD)	12.29 (3.21)		11.99 (3.03)		
Depression					0.163
Absence (<7)	0	0.00	3	2.00	
Suspected (8–10)	55	35.03	57	38.00	
Confirmed (>10)	102	64.97	90	60.00	
Mean depression (SD)	20.25 (3.26)		20.46 (3.22)		
End of radiotherapy (T2)					
Anxiety					
Absence (<7)	0	0.00	0	0.00	
Suspected (8–10)	0	0.00	0	0.00	
Confirmed (>10)	147	100.00	142	100.00	
Mean anxiety (SD)	11.69 (3.04)		11.97 (3.04)		
Depression					0.576
Absence (<7)	1	0.68	1	0.70	
Suspected (8–10)	68	46.26	57	40.14	
Confirmed (>10)	78	53.06	84	59.15	
Mean depression (SD)	20.38 (3.06)		20.08 (3.34)		
1 year after inclusion (T3)					
Anxiety					0.367
Absence (<7)	0	0.00	1	0.81	
Suspected (8–10)	1	0.77	0	0.00	
Confirmed (>10)	129	99.23	123	99.19	
Mean anxiety (SD)	11.85 (2.42)		11.81 (3.11)		
Depression					0.052
Absence (<7)	0	0.00	2	1.61	
Suspected (8–10)	43	33.08	55	44.35	
Confirmed (>10)	87	66.92	67	54.03	
Mean depression (SD)	20.26 (2.87)		19.87 (3.62)		

3.3. Physical Activity

Dimensions of the GPAQ and sit-to-stand test are described in Table 4.

Table 4. Physical activity according to the GPAQ and muscular test in the intention-to-treat population.

		Baseline (T0) (158/168)		End of CT (T1) (n = 139; n = 129)			End of RT (T2) (n = 134; n = 128)			1 Year after Inclusion (T3) (n = 113; n = 97)			LMM Coefficients [1] (95% CI)
		Mean	SD	Mean	SD	p	Mean	SD	p	Mean	SD	p	
Total MET (MET.min/wk)	Control	1998.63	2632.06	2339.37	4181.16	0.03	1522.69	1591.27	0.048	2157.1	2902.02	0.03	$\beta_1 = 0.74$ [0.37; 1.10], $p < 0.001$
	APAD	2133.07	3206.8	2023.1	1729.05		1760.78	1464.05		2363.96	2386.1		$\beta_0 = -0.005$ [−0.007; 0.007], $p = 0.864$
Recreational–Moderate intensity (MET.min/wk)	Control	338	511.97	299.08	539.7	<0.001	395.82	538.71	0.001	506.27	728.95	0.349	$\beta_1 = 0.95$ [0.46; 1.43], $p < 0.001$
	APAD	296.98	454.64	632.87	730.3		600.56	630.25		537.2	694.62		$\beta_0 = 0.021$ [0.012; 0.030], $p < 0.001$
Recreational–Vigorous intensity (MET.min/wk)	Control	122.78	461.56	219.86	829.16	0.433	173.73	618.49	0.636	216.28	777.94	0.353	$\beta_1 = 0.27$ [−0.13; 0.67], $p = 0.184$
	APAD	135	450.99	187.6	538.19		220.31	767.45		274.06	654.44		$\beta_0 = 0.008$ [0.001; 0.015], $p = 0.019$
Work—Moderate intensity (MET.min/wk)	Control	816.46	1479.06	515.11	1419.7	0.006	429.28	849.11	0.674	667.65	1423.57	0.743	$\beta_1 = 0.34$ [−0.19; 0.86], $p = 0.212$
	APAD	824.4	1638.35	576.59	1001.46		508.78	998.82		546.1	963.79		$\beta_0 = -0.012$ [0.022; −0.002], $p = 0.014$
Work—Vigorous intensity (MET.min/wk)	Control	212.41	1378.38	523.17	2778.69	0.935	53.73	316.49	0.813	120.35	804.99	0.721	$\beta_1 = 0.10$ [−0.16; 0.36], $p = 0.454$
	APAD	374.29	1906.61	78.76	420.47		79.38	556.03		225.15	1119.72		$\beta_0 = 0.00003$ [−0.005; 0.005], $p = 0.989$
Travel—Moderate intensity (MET.min/wk)	Control	508.99	816.59	782.16	1117.32	0.071	470.12	702.81	0.277	646.55	1307.34	0.141	$\beta_1 = 0.19$ [−0.35; 0.72], $p = 0.494$
	APAD	502.4	805.68	547.29	964.76		465.23	542.61		781.44	1734.93		$\beta_0 = 0.01$ [0.01; 0.02], $p = 0.022$
Sitting or reclining time (min/day)	Control	372.41	175.26	355.29	197.46	0.041	357.07	171.3	0.84	337.12	164.13	0.946	$\beta_1 = 0.18$ [0.02; 0.33], $p = 0.023$
	APAD	400.85	178.62	391.05	177.33		369.2	177.26		353.76	183.5		$\beta_0 = -0.004$ [−0.008; −0.0004], $p = 0.029$
Muscular test													
Sit-to-stand 30 s	Control	18.09	4.67	17.3	5.51	0.013	18.86	5.32	0.39	19.42	6.02	0.615	$\beta_1 = 0.001$ [−0.002; 0.002], $p = 0.094$ $\beta_0 = 0.001$ [0.0002; 0.002], $p = 0.017$
	APAD	17.72	4.87	18.61	4.77		19.68	5.64		20.04	6.16		
Sit-to-stand ratio (30 s/15 s)	Control	1.97	0.38	1.93	0.2	0.341	1.94	0.16	0.59	1.93	0.16	0.844	$\beta_1 = -0.002$ [−0.014; 0.01], $p = 0.694$ $\beta_0 = -0.0001$ [−0.0004; 0.0002], $p = 0.431$
	APAD	1.93	0.18	1.94	0.17		1.95	0.13		1.93	0.15		

[1] In the linear mixed model (LMM; log transformed variables): β_1 is the coefficient of the variable 'arm' (interpreted as APAD effect with respect to Control) noted as β_2 when interaction arm by time was significant. β_0 is the coefficient of the variable time (weeks) variable. Note: No baseline imbalance was observed for the studied variables (for sitting or reclining time (GPAQ), $p = 0.136$). CT: chemotherapy, RT: radiotherapy.

The intervention had a positive impact on the total MET (Figure 4) and on the moderate intensity recreational activities (Figure 5), as they were significantly higher in the APAD arm with respect to control ($\beta_1 = 0.74$ [0.37; 1.10]; $p < 0.001$ and $\beta_1 = 0.96$ [0.46; 1.43]; $p < 0.001$). However, the sitting or reclining time per day appeared to be slightly higher in the APAD arm ($\beta_1 = 0.18$ [0.02; 0.33]; $p = 0.023$; Figure S5).

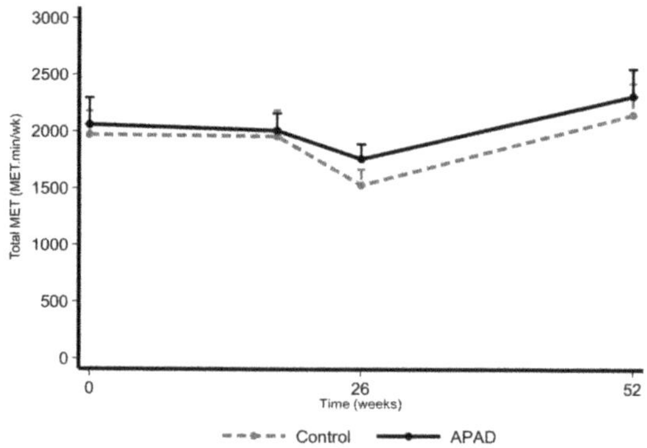

Figure 4. Evolution of the total MET based on the GPAQ according to randomization arm in the intention-to-treat population. Data are presented as mean + SD.

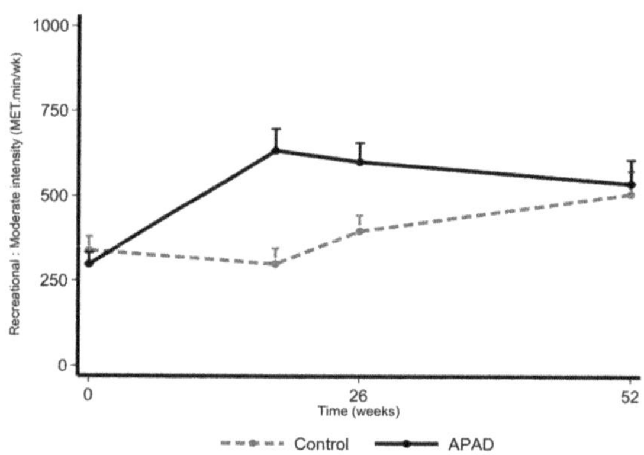

Figure 5. Moderate intensity recreational MET over time based on the GPAQ according to randomization arm in the intention-to-treat population. Data are presented as mean + SD.

Regarding compliance with WHO recommendations (Table 5), patients in the APAD arm more frequently met the recommendations compared to the control arm at T1 (81.40% vs. 61.87%, $p < 0.001$) and T3 (86.60% vs. 68.14%, $p = 0.002$).

Table 5. Compliance with the WHO Stepwise [1] recommendations for physical activity (GPAQ) in the intention-to-treat population.

		Control		APAD		Total		
		n	%	*n*	%	*n*	%	*p*
Baseline (T0)								
Low activity								0.084
	No	105	66.46	96	57.14	201	61.66	
	Yes	53	33.54	72	42.86	125	38.34	
Failed to meet WHO recommendations								0.568
	No	118	74.68	130	77.38	248	76.07	
	Yes	40	25.32	38	22.62	78	23.93	
End of chemotherapy (T1)								
Low activity								0.004
	No	71	51.08	88	68.22	159	59.33	
	Yes	68	48.92	41	31.78	109	40.67	
Failed to meet WHO recommendations								0.000
	No	86	61.87	105	81.40	191	71.27	
	Yes	53	38.13	24	18.60	77	28.73	
End of radiotherapy (T2)								
Low activity								0.094
	No	81	60.45	90	70.31	171	65.27	
	Yes	53	39.55	38	29.69	91	34.73	
Failed to meet WHO recommendations								0.071
	No	95	70.90	103	80.47	198	75.57	
	Yes	39	29.10	25	19.53	64	24.43	
1 year after inclusion (T3)								
Low activity								0.003
	No	64	56.64	74	76.29	138	65.71	
	Yes	49	43.36	23	23.71	72	34.29	
Failed to meet WHO recommendations								0.002
	No	77	68.14	84	86.60	161	76.67	
	Yes	36	31.86	13	13.40	49	23.33	

[1] STEPS analysis program.

3.4. Dietary Intake

The evolution of dietary intake and weight control by randomization arm is summarized in Table 6 and Figure S6. At T2, a significant increase in the consumption of fiber was observed in the APAD arm ($p = 0.020$), as well as a reduction in the consumption of animal proteins and alcohol ($p = 0.003$ and $p = 0.019$, respectively). However, both reductions seem to have been only temporary, as 6 months later the effect was diminished. These results were confirmed when we analyzed the evolution of these parameters over time in an LMM.

According to the adjusted LMM, the intervention had a positive significant impact on increasing fiber consumption over time ($\beta_1 = 0.096$ [0.026; 0.17]; $p = 0.007$). A trend of reduced consumption of animal proteins was also observed and was lower in the APAD arm than the control arm ($\beta_1 = -0.26$ [−0.55; 0.017]; $p = 0.066$). No change in weight over time was observed.

Table 6. Dietary intake per day (3-day record) and weight control variables in the intention-to-treat population.

		Baseline (T0)			End of Radiotherapy (T2)			1 Year after Inclusion (T3)			LMM Coefficients [1] [95% CI]
		Mean	SD	p	Mean	SD	p	Mean	SD	p	
Total energy (Kcal)	Control	1507.46	362.87	0.312	1402.41	386.27	0.637	1382.97	384.02	0.16	$\beta_1 = -0.0020$ [−0.054; 0.050], $p = 0.947$
	APAD	1463.76	434.09		1377.56	373.81		1452.03	372.93		$\beta_0 = -0.0010$ [−0.002; −0.0002], $p = 0.013$
Animal proteins (g)	Control	19.64	17.92	0.233	10.38	10.04	0.003	8.35	8.99	0.233	$\beta_1 = -0.26$ [−0.55; 0.017], $p = 0.066$
	APAD	17.29	17.03		7.95	9.94		7.42	8.55		$\beta_0 = -0.017$ [−0.022; −0.011], $p < 0.001$
Vegetal proteins (g)	Control	8.85	5.93	1	5.61	4.29	0.415	5.49	4.3	0.836	$\beta_1 = -0.084$ [−0.23; 0.067], $p = 0.276$
	APAD	9.2	6.87		5.45	5.59		5.53	5.07		$\beta_0 = -0.010$ [−0.013; −0.007], $p < 0.01$
Lipids (g)	Control	61.43	21.98	0.147	57	22.9	0.22	54.82	24.05	0.407	$\beta_1 = -0.037$ [−0.11; 0.034], $p = 0.306$
	APAD	57.72	23.49		52.32	18.73		56.34	19.54		$\beta_0 = -0.001$ [−0.003; −0.0002], $p = 0.021$
Monounsaturated lipids (g)	Control	22	9.64	0.207	19.39	9.95	0.805	19.73	12.88	0.538	$\beta_1 = -0.024$ [−0.11; 0.057], $p = 0.558$
	APAD	20.55	9.27		18.37	7.15		19.62	7.51		$\beta_0 = -0.002$ [−0.003; 0.0004], $p = 0.009$
Polyunsaturated lipids (g)	Control	7.86	5.35	0.549	7.92	5.32	0.791	7.26	4.51	0.085	$\beta_1 = 0.022$ [−0.078; 0.12], $p = 0.671$
	APAD	7.4	4.31		7.35	3.97		8.11	4.42		$\beta_0 = 0.0002$ [−0.001; 0.002], $p = 0.773$
Simple sugars (g)	Control	69.97	28.08	0.926	61.93	23.26	0.273	61.26	24.46	0.504	$\beta_1 = 0.023$ [−0.058; 0.10], $p = 0.582$
	APAD	70.25	30.15		65.66	24.96		64.03	26.17		$\beta_0 = -0.002$ [−0.004; 0.001], $p < 0.001$
Alcohol (g)	Control	4.18	6.58	0.742	4.11	5.95	0.019	4.15	6.83	0.055	$\beta_1 = -0.091$ [−0.010; −0.0009], $p = 0.549$
	APAD	4.44	7.31		2.25	4.18		2.98	5.67		$\beta_0 = -0.005$ [−0.010; -0.0009], $p = 0.020$
Fiber (g)	Control	15.54	5.56	0.277	15.35	5.73	0.02	15.35	6.03	0.003	$\beta_1 = 0.096$ [0.026; 0.17], $p = 0.007$
	APAD	16.28	5.87		17.37	6.49		17.58	5.9		$\beta_0 = 0.0005$ [−0.0006; 0.001], $p = 0.389$
Weight control											
Weight (kg)	Control	67	14.13	0.44	66.23	12.93	0.571	67.13	13.98	0.576	$\beta_1 = 0.020$ [−0.020; 0.060], $p = 0.334$
	APAD	68.41	14.6		67.57	13.56		68.28	14.19		$\beta_0 = 0.0001$ [−0.0001; 0.0002], $p = 0.389$
BMI (kg/m^2)	Control	25.22	5.3	0.317	24.95	5.01	0.51	25.23	5.28	0.655	$\beta_1 = 0.020$ [−0.019; 0.059], $p = 0.320$
	APAD	25.72	5.14		25.29	4.79		25.47	5.01		$\beta_0 = 0.0001$ [−0.0001; 0.0002], $p = 0.337$
Waist size (cm)	Control	163.08	6.49	0.717	87.09	13.05	0.827	87.26	15.74	0.344	$\beta_1 = 0.0034$ [−0.027; 0.034], $p = 0.827$
	APAD	163.06	6.32		86.87	11.72		88.65	13.48		$\beta_0 = 0.0002$ [−0.0001; 0.0004], $p = 0.133$

[1] In the linear mixed model (LMM; log transformed variables): β_1 is the coefficient of the variable 'arm' (interpreted as APAD effect with respect to Control) noted as β_2 when interaction arm by time was significant. β_0 is the coefficient of the variable time (weeks) variable. Note: No baseline imbalance was observed for the studied variables.

3.5. Chemotherapy Completion Rates

The RDI was high overall, with 90.7% and 80% of the patients having an RDI > 80% and 90%, respectively, in the intention-to-treat population. No difference was found between the two arms (80.79% vs. 79.21%, $p = 0.71$ for RDI > 90%; 91.53% vs. 89.89%, $p = 0.595$ for RDI > 80% in APAD and control arms, respectively).

4. Discussion

In the present study, we reported the results of a large, multicenter, randomized trial evaluating the impact of an APAD education program on fatigue, fat mass, and health-related QoL. The results do not support our previous conclusions [18,20] and the main hypothesis of the efficacy of the exercise and diet education program to relieve cancer-related fatigue in EBC patients receiving adjuvant treatments. One of the main pitfalls could be linked to patient adherence to the protocol. The drop-out rate in the present study was higher in the APAD arm, with 17% of the population discontinuing the program early. In addition, the intensity of physical activity appears to have insufficient autonomy, with low adherence to the intervention. Twenty-one percent of the patients did not perform the complete supervised exercise sessions, and only 65% had completed at least 80% of the home sessions. Our intervention appears to change low activity only. Thus, this education program appears to be more of a lifestyle change, in terms of adherence and clinical impact, to more active practice, as confirmed by the high proportion of patients achieving the WHO targets without achieving a level of physical activity high enough to induce a clinically significant reduction in fatigue. The design of our program, with discontinuous supervision, may explain the ambiguous results for fatigue during follow-up [40–45].

As reported in recent meta-analyses and guidelines, supervised exercise interventions appear to have significantly greater effects on fatigue than unsupervised exercise interventions, and shorter supervised interventions with a duration ≤12 weeks appear to induce greater effects on fatigue than supervised interventions with a longer duration [9,46]. However, the results indicate significant benefits on depressive symptoms. This impact supports a reported decrease in psychological distress and an improvement in self-esteem with exercise [9].

In contrast to our hypothesis, we found no significant effect of the intervention on BMI and the chemotherapy completion rates. Previous reports on combined diet and exercise interventions delivered during chemotherapy [16,17] failed to find any benefits on hip circumference and BMI. One explanation could be based on the significant, but transient, modification of nutritional intake, specifically animal proteins, alcohol, and fiber, without persistent marked behavioral modifications, as highlighted by the trends in protein and lipid consumption in the LMM. Evaluation of the impact of continued education programs, with later-time supervised sessions, could help define the best method of sustaining early changes.

A meta-analysis of 32 randomized studies comprising 2626 EBC patients evaluated the impact of supervised aerobic or resistance exercise interventions during adjuvant treatment and reported a pooled significant improvement in strength [47]. High-dose training or a focus on resistance training appears to be associated with better effects on strength in this patient population [41,45,48]. At the same time, health education interventions appear to be associated with a lesser impact on cancer-related fatigue compared to exercise training [49].

Regarding behavioral outcomes after combined interventions including diet and exercise components delivered during chemotherapy, two studies [16,17] yielded significant changes in dietary intake and one presented significant changes in total physical activity in the intervention group (post- vs. pre-intervention) [17]. Our APAD intervention had a significantly favorable impact on leisure time physical activity at the end of chemotherapy and the end of radiotherapy, but improvements in total physical activity were not significant. Physical activity done in the framework of the APAD intervention was reported in the leisure category, which explains the enhancement of that physical activity type in our study. Regarding dietary intake, no significant changes were observed in the APAD arm of our study versus usual care. The 3-day record method and food to nutrition conversion software we used generated large standard deviations (see Table 4, baseline values for nutrients) that possibly impaired the statistical power to detect a between-group difference. The two previous studies [16,17] that demonstrated dietary changes analyzed dietary data that were collected by food frequency questionnaires.

Few studies evaluating a diet and/or exercise intervention have included follow-up measures [40–44]. Most of them, such as our present study, failed to show that significant effects were maintained after the end of the intervention [40,42,44,45]. One study reported improvements in the 6-min walk test 6 months post-intervention [43]. Difficulties could be related to the weak supervision and support over time, making patients unable to maintain the initially induced changes in behavior. The impact of the intervention seems to be limited in time, even for trials evaluating different doses of exercise in this context [50]. This finding may promote the necessity of setting longer supervised intervention models that can include, for example, the present "in-treatment" module during the chemotherapy part of the treatment plan, followed by additional supervised sessions during radiotherapy and a 6-month internet-based "survivor" module designed to maintain behavioral changes and support autonomy with limited cost. Several telephone- or internet-based diet–exercise interventions have been tested in BC survivors and yielded health benefits [51–53] with moderate to good adherence rates (from 41–87% of adherent patients) [54]. However, these adherence rates pinpoint the need for enduring strategies to maintain the motivation of this "survivor" population. In the study by Stone et al., adherence was significantly associated with moderate levels of exercise interventions (similar to the activity levels selected in our present study), BMI, physical health, and employment status, with reduced adherence in women working full-time compared to those who do not work full-time [55]. These findings, combined with the difference in general fatigue found in the present study, encompass the impact of social inequalities on the global impact of cancer and cancer treatments, as well as access and adherence to supportive care programs in these populations.

Our study was designed primarily to validate our previous results by assessing, in a pragmatic context, the effectiveness of an education program on exercise and diet compared to the standard of care in France (the usual care control arm). Though pragmatic and comprehensive, the main drawback of this design is that it does not allow disentanglement of the independent effects of exercise and dietary components. Another limitation is the differential drop-out rate in the APAD vs. usual care groups. Although many outcomes improved significantly in the APAD arm, and an increase was observed in the declared leisure physical activity, the between-group difference was not significant for objectively measured physical activity. The spontaneous physical activity level of the usual care group may have partly diluted the effects of the APAD intervention. Given the number of comparisons we made at each time point for the secondary outcomes without adjusting for multiple testing, we would expect a few false discoveries by chance.

Finally, based on the publication of Smets et al. [31], we hypothesized that our population would be affected by significant levels of cancer- and treatment-induced fatigue. However, our population was affected by much lower levels of fatigue than the expected mean Global fatigue score of 16, as the mean General fatigue scores were 9.95 and 9.63 for the APAD and control arms, respectively. These relatively low levels of fatigue could explain some of the negative results. Buffart et al. recently reported in a meta-analysis that, even if exercise induced a benefit for all patients during treatment, only patients affected by the worse levels of fatigue experienced a persistent impact of exercise post-treatment [56]. Thus, targeting specific subgroups of patients with higher scores for fatigue or variables associated with a greater level of fatigue during treatment may be more beneficial and cost-effective. In this regard, the observation of a gradient in the level of fatigue throughout the treatment period according to the stratum of precariousness confirms that the level of precariousness evaluated using the EPICES score is a stratification factor and fully justifies its inclusion as a stratification factor in this study. It seems important to consider this social parameter in the design of future studies evaluating the impact of supportive care interventions during the treatment of EBC.

5. Conclusions

In French patients receiving adjuvant treatment for EBC, a mixed hospital- and home-based diet and exercise education program during adjuvant chemotherapy and radiation therapy inducing a transient modification in the level of physical activity and dietary intake is not sufficient to demonstrate improvements in fatigue during and after treatment. We found no impact on the dose intensity of chemotherapy. Fatigue appears to correlate with the level of precariousness. Cancer care centers should consider integrating more proactive diet–exercise supportive care into the management of patients with BC who are receiving chemotherapy and/or radiotherapy, focusing on fatigued and/or precarious patients. Additional information is needed in order to identify the optimal intervention timing to induce persistent lifestyle changes and reduce long term alterations of patients' quality of life in this setting.

Supplementary Materials: The following are available online at http://www.mdpi.com/2072-6643/12/10/3081/s1, Figure S1: General Fatigue subscale of the MFI20 according to EPICE strata (precariousness level), Figure S2: General Fatigue subscale of the MFI20 according to randomization arms stratified by EPICE score (precariousness level), Figure S3: Evolution of quality of life (EORTC QLQ-C30) in the intention-to-treat population, Figure S4: Evolution of quality of life (EORTC QLQ-C30) in the per protocol population, Figure S5: Evolution of the different physical activity domains of the GPAQ questionnaire in the intention-to-treat population, Figure S6: Evolution of nutrient intake and weight control variables (weight, BMI, waist size) by randomization arm in the intention-to-treat population, Table S1: Fatigue sub-scales of the MFI20 over time in the per protocol population, Table S2: Anxiety and depression disorders in the per protocol population.

Author Contributions: Conceptualization, W.J., P.S., M.C., J.-P.B., S.G., G.R. and G.N.; Data curation, M.J., S.G., C.J. and S.L.; Formal analysis, W.J. and M.J.; Funding acquisition, J.-P.B. and G.R.; Investigation, W.J., A.A. (Antoine Arnaud), C.L.-P., P.D., A.A. (Ahmed Azzedine), O.T., S.S.-L., S.M., V.D., G.L., J.G., G.R. and L.V.; Methodology, M.J., P.S., M.C., J.-P.B., S.G., G.R. and G.N.; Project administration, J.-P.B., C.J., S.L. and G.R.; Resources, J.-P.B.; Supervision, W.J., S.M., J.-P.B., S.G., C.J., G.R. and G.N.; Validation, W.J., A.A. (Antoine Arnaud), M.J., C.L.-P., P.D., P.S., A.A. (Ahmed Azzedine), O.T., S.S.-L., S.M., M.C., J.-P.B., S.G., C.J., S.L., V.D., G.L., J.G., G.R., G.N. and L.V.; Writing–original draft, W.J., A.A. (Antoine Arnaud), M.J., C.L.-P., P.D., P.S., A.A. (Ahmed Azzedine), O.T., S.S.-L., S.M., M.C., J.-P.B., S.G., C.J., S.L., V.D., G.L., J.G., G.R., G.N. and L.V.; Writing–review & editing, W.J., A.A. (Antoine Arnaud), M.J., C.L.-P., P.D., P.S., A.A. (Antoine Arnaud), O.T., S.S.-L., S.M., M.C., J.-P.B., S.G., C.J., S.L., V.D., G.L., J.G., G.R., G.N. and L.V. All authors have read and agreed to the published version of the manuscript.

Funding: This research was funded by a grant from the INCa-DGOS (Grant SHP-ESF 2012-079) and by a Montpellier Cancer SIRIC grant (INCa_Inserm_DGOS_12553).

Conflicts of Interest: The authors declare no conflict of interest. The funders had no role in the design of the study; in the collection, analyses, or interpretation of data; in the writing of the manuscript, or in the decision to publish the results.

Abbreviations

APAD	adapted physical activity and diet
EBC	early breast cancer
BMI	body mass index
GPAQ	Global Physical Activity Questionnaire
MFI	Multidimensional Fatigue Inventory
PRO	patient-reported outcome
QoL	quality-of-life
RCT	randomized controlled trial
RDI	relative dose index
SD	standard deviation

References

1. Henry, D.H.; Viswanathan, H.N.; Elkin, E.P.; Traina, S.; Wade, S.; Cella, D. Symptoms and treatment burden associated with cancer treatment: Results from a cross-sectional national survey in the U.S. *Support. Care Cancer Off. J. Multinatl. Assoc. Support. Care Cancer* **2008**, *16*, 791–801. [CrossRef] [PubMed]

2. Fabi, A.; Falcicchio, C.; Giannarelli, D.; Maggi, G.; Cognetti, F.; Pugliese, P. The course of cancer related fatigue up to ten years in early breast cancer patients: What impact in clinical practice? *Breast* **2017**, *34*, 44–52. [CrossRef] [PubMed]
3. Williams, L.A.; Bohac, C.; Hunter, S.; Cella, D. Patient and health care provider perceptions of cancer-related fatigue and pain. *Support. Care Cancer Off. J. Multinatl. Assoc. Support. Care Cancer* **2016**, *24*, 4357–4363. [CrossRef] [PubMed]
4. Vogelzang, N.J.; Breitbart, W.; Cella, D.; Curt, G.A.; Groopman, J.E.; Horning, S.J.; Itri, L.M.; Johnson, D.H.; Scherr, S.L.; Portenoy, R.K. Patient, caregiver, and oncologist perceptions of cancer-related fatigue: Results of a tripart assessment survey. The Fatigue Coalition. *Semin. Hematol.* **1997**, *34*, 4–12. [PubMed]
5. Penttinen, H.M.; Saarto, T.; Kellokumpu-Lehtinen, P.; Blomqvist, C.; Huovinen, R.; Kautiainen, H.; Jarvenpaa, S.; Nikander, R.; Idman, I.; Luoto, R.; et al. Quality of life and physical performance and activity of breast cancer patients after adjuvant treatments. *Psycho Oncol.* **2011**, *20*, 1211–1220. [CrossRef]
6. Montazeri, A. Quality of life data as prognostic indicators of survival in cancer patients: An overview of the literature from 1982 to 2008. *Health Qual. Life Outcomes* **2009**, *7*, 102. [CrossRef]
7. Chan, D.S.; Vieira, A.R.; Aune, D.; Bandera, E.V.; Greenwood, D.C.; McTiernan, A.; Navarro Rosenblatt, D.; Thune, I.; Vieira, R.; Norat, T. Body mass index and survival in women with breast cancer-systematic literature review and meta-analysis of 82 follow-up studies. *Ann. Oncol. Off. J. Eur. Soc. Med Oncol.* **2014**, *25*, 1901–1914. [CrossRef]
8. Playdon, M.C.; Bracken, M.B.; Sanft, T.B.; Ligibel, J.A.; Harrigan, M.; Irwin, M.L. Weight Gain After Breast Cancer Diagnosis and All-Cause Mortality: Systematic Review and Meta-Analysis. *J. Natl. Cancer Inst.* **2015**, *107*, djv275. [CrossRef]
9. Campbell, K.L.; Winters-Stone, K.M.; Wiskemann, J.; May, A.M.; Schwartz, A.L.; Courneya, K.S.; Zucker, D.S.; Matthews, C.E.; Ligibel, J.A.; Gerber, L.H.; et al. Exercise Guidelines for Cancer Survivors: Consensus Statement from International Multidisciplinary Roundtable. *Med. Sci. Sports Exerc.* **2019**, *51*, 2375–2390. [CrossRef]
10. Berger, A.M.; Mooney, K.; Alvarez-Perez, A.; Breitbart, W.S.; Carpenter, K.M.; Cella, D.; Cleeland, C.; Dotan, E.; Eisenberger, M.A.; Escalante, C.P.; et al. Cancer-Related Fatigue, Version 2.2015. *J. Natl. Compr. Cancer Netw. JNCCN* **2015**, *13*, 1012–1039. [CrossRef]
11. Jones, L.W.; Demark-Wahnefried, W. Diet, exercise, and complementary therapies after primary treatment for cancer. *Lancet Oncol.* **2006**, *7*, 1017–1026. [CrossRef]
12. Denlinger, C.S.; Ligibel, J.A.; Are, M.; Baker, K.S.; Demark-Wahnefried, W.; Dizon, D.; Friedman, D.L.; Goldman, M.; Jones, L.; King, A.; et al. Survivorship: Nutrition and weight management, Version 2.2014. Clinical practice guidelines in oncology. *J. Natl. Compr. Cancer Netw.* **2014**, *12*, 1396–1406. [CrossRef]
13. Demark-Wahnefried, W.; Morey, M.C.; Sloane, R.; Snyder, D.C.; Miller, P.E.; Hartman, T.J.; Cohen, H.J. Reach out to enhance wellness home-based diet-exercise intervention promotes reproducible and sustainable long-term improvements in health behaviors, body weight, and physical functioning in older, overweight/obese cancer survivors. *J. Clin. Oncol. Off. J. Am. Soc. Clin. Oncol.* **2012**, *30*, 2354–2361. [CrossRef] [PubMed]
14. Goodwin, P.J.; Segal, R.J.; Vallis, M.; Ligibel, J.A.; Pond, G.R.; Robidoux, A.; Blackburn, G.L.; Findlay, B.; Gralow, J.R.; Mukherjee, S.; et al. Randomized trial of a telephone-based weight loss intervention in postmenopausal women with breast cancer receiving letrozole: The LISA trial. *J. Clin. Oncol. Off. J. Am. Soc. Clin. Oncol.* **2014**, *32*, 2231–2239. [CrossRef] [PubMed]
15. Rock, C.L.; Flatt, S.W.; Byers, T.E.; Colditz, G.A.; Demark-Wahnefried, W.; Ganz, P.A.; Wolin, K.Y.; Elias, A.; Krontiras, H.; Liu, J.; et al. Results of the Exercise and Nutrition to Enhance Recovery and Good Health for You (ENERGY) Trial: A Behavioral Weight Loss Intervention in Overweight or Obese Breast Cancer Survivors. *J. Clin. Oncol. Off. J. Am. Soc. Clin. Oncol.* **2015**, *33*, 3169–3176. [CrossRef] [PubMed]
16. Demark-Wahnefried, W.; Case, L.D.; Blackwell, K.; Marcom, P.K.; Kraus, W.; Aziz, N.; Snyder, D.C.; Giguere, J.K.; Shaw, E. Results of a diet/exercise feasibility trial to prevent adverse body composition change in breast cancer patients on adjuvant chemotherapy. *Clin. Breast Cancer* **2008**, *8*, 70–79. [CrossRef] [PubMed]
17. Djuric, Z.; Ellsworth, J.S.; Weldon, A.L.; Ren, J.; Richardson, C.R.; Resnicow, K.; Newman, L.A.; Hayes, D.F.; Sen, A. A Diet and Exercise Intervention during Chemotherapy for Breast Cancer. *Open Obes. J.* **2011**, *3*, 87–97. [CrossRef]

18. Carayol, M.; Ninot, G.; Senesse, P.; Bleuse, J.P.; Gourgou, S.; Sancho-Garnier, H.; Sari, C.; Romieu, I.; Romieu, G.; Jacot, W. Short- and long-term impact of adapted physical activity and diet counseling during adjuvant breast cancer therapy: The "APAD1" randomized controlled trial. *BMC Cancer* **2019**, *19*, 737. [CrossRef]
19. Perrier, L.; Foucaut, A.M.; Morelle, M.; Touillaud, M.; Kempf-Lepine, A.S.; Heinz, D.; Gomez, F.; Meyrand, R.; Baudinet, C.; Berthouze, S.; et al. Cost-effectiveness of an exercise and nutritional intervention versus usual nutritional care during adjuvant treatment for localized breast cancer: The PASAPAS randomized controlled trial. *Support. Care Cancer Off. J. Multinatl. Assoc. Support. Care Cancer* **2020**, *28*, 2829–2842. [CrossRef]
20. Carayol, M.; Romieu, G.; Bleuse, J.P.; Senesse, P.; Gourgou-Bourgade, S.; Sari, C.; Jacot, W.; Sancho-Garnier, H.; Janiszewski, C.; Launay, S.; et al. Adapted physical activity and diet (APAD) during adjuvant breast cancer therapy: Design and implementation of a prospective randomized controlled trial. *Contemp. Clin. Trials* **2013**, *36*, 531–543. [CrossRef]
21. Eisinger, F.; Viguier, J.; Touboul, C.; Coscas, Y.; Pivot, X.; Blay, J.Y.; Lhomel, C.; Morere, J.F. Social stratification, risk factor prevalence and cancer screening attendance. *Eur. J. Cancer Prev. Off. J. Eur. Cancer Prev. Organ.* **2015**, *24*, S77–S81. [CrossRef]
22. Labbe, E.; Blanquet, M.; Gerbaud, L.; Poirier, G.; Sass, C.; Vendittelli, F.; Moulin, J.J. A new reliable index to measure individual deprivation: The EPICES score. *Eur. J. Public Health* **2015**, *25*, 604–609. [CrossRef] [PubMed]
23. Morere, J.F.; Eisinger, F.; Touboul, C.; Lhomel, C.; Couraud, S.; Viguier, J. Decline in Cancer Screening in Vulnerable Populations? Results of the EDIFICE Surveys. *Curr. Oncol. Rep.* **2018**, *20*, 17. [CrossRef] [PubMed]
24. Roche, H.; Fumoleau, P.; Spielmann, M.; Canon, J.L.; Delozier, T.; Serin, D.; Symann, M.; Kerbrat, P.; Soulie, P.; Eichler, F.; et al. Sequential adjuvant epirubicin-based and docetaxel chemotherapy for node-positive breast cancer patients: The FNCLCC PACS 01 Trial. *J. Clin. Oncol. Off. J. Am. Soc. Clin. Oncol.* **2006**, *24*, 5664–5671. [CrossRef]
25. Clinton, S.K.; Giovannucci, E.L.; Hursting, S.D. The World Cancer Research Fund/American Institute for Cancer Research Third Expert Report on Diet, Nutrition, Physical Activity, and Cancer: Impact and Future Directions. *J. Nutr.* **2020**, *150*, 663–671. [CrossRef] [PubMed]
26. Schmitz, K.H.; Courneya, K.S.; Matthews, C.; Demark-Wahnefried, W.; Galvao, D.A.; Pinto, B.M.; Irwin, M.L.; Wolin, K.Y.; Segal, R.J.; Lucia, A.; et al. American College of Sports Medicine roundtable on exercise guidelines for cancer survivors. *Med. Sci. Sports Exerc.* **2010**, *42*, 1409–1426. [CrossRef]
27. Courneya, K.S.; Mackey, J.R.; McKenzie, D.C. Exercise for breast cancer survivors: Research evidence and clinical guidelines. *Physician Sportsmed.* **2002**, *30*, 33–42. [CrossRef] [PubMed]
28. Wolin, K.Y.; Schwartz, A.L.; Matthews, C.E.; Courneya, K.S.; Schmitz, K.H. Implementing the exercise guidelines for cancer survivors. *J. Supportive Oncol.* **2012**, *10*, 171–177. [CrossRef]
29. Roza, A.M.; Shizgal, H.M. The Harris Benedict equation reevaluated: Resting energy requirements and the body cell mass. *Am. J. Clin. Nutr.* **1984**, *40*, 168–182. [CrossRef]
30. Martin, A.; Touvier, M.; Volatier, J.L. The basis for setting the upper range of adequate intake for regulation of macronutrient intakes, especially amino acids. *J. Nutr.* **2004**, *134*, 1625S–1629S, discussion 1630S–1632S, 1667S–1672S. [CrossRef] [PubMed]
31. Smets, E.M.; Garssen, B.; Cull, A.; de Haes, J.C. Application of the multidimensional fatigue inventory (MFI-20) in cancer patients receiving radiotherapy. *Br. J. Cancer* **1996**, *73*, 241–245. [CrossRef] [PubMed]
32. Gentile, S.; Delaroziere, J.C.; Favre, F.; Sambuc, R.; San Marco, J.L. Validation of the French 'multidimensional fatigue inventory' (MFI 20). *Eur. J. Cancer Care* **2003**, *12*, 58–64. [CrossRef] [PubMed]
33. Razavi, D.; Delvaux, N.; Farvacques, C.; Robaye, E. Screening for adjustment disorders and major depressive disorders in cancer in-patients. *Br. J. Psychiatry J. Ment. Sci.* **1990**, *156*, 79–83. [CrossRef] [PubMed]
34. Sprangers, M.A.; Cull, A.; Bjordal, K.; Groenvold, M.; Aaronson, N.K. The European Organization for Research and Treatment of Cancer. Approach to quality of life assessment: Guidelines for developing questionnaire modules. EORTC Study Group on Quality of Life. *Qual. Life Res. Int. J. Qual. Life Asp. Treat. Care Rehabil.* **1993**, *2*, 287–295. [CrossRef] [PubMed]
35. Trinh, O.T.; Nguyen, N.D.; van der Ploeg, H.P.; Dibley, M.J.; Bauman, A. Test-retest repeatability and relative validity of the Global Physical Activity Questionnaire in a developing country context. *J. Phys. Act. Health* **2009**, *6*, S46–S53. [CrossRef] [PubMed]

36. Ruiz-Casado, A.; Alejo, L.B.; Santos-Lozano, A.; Soria, A.; Ortega, M.J.; Pagola, I.; Fiuza-Luces, C.; Palomo, I.; Garatachea, N.; Cebolla, H.; et al. Validity of the Physical Activity Questionnaires IPAQ-SF and GPAQ for Cancer Survivors: Insights from a Spanish Cohort. *Int. J. Sports Med.* **2016**, *37*, 979–985. [CrossRef]
37. Bull, F.C.; Maslin, T.S.; Armstrong, T. Global physical activity questionnaire (GPAQ): Nine country reliability and validity study. *J. Phys. Act. Health* **2009**, *6*, 790–804. [CrossRef]
38. Biro, G.; Hulshof, K.F.; Ovesen, L.; Amorim Cruz, J.A.; Group, E. Selection of methodology to assess food intake. *Eur. J. Clin. Nutr.* **2002**, *56*, S25–S32. [CrossRef]
39. Aaronson, N.K.; Ahmedzai, S.; Bergman, B.; Bullinger, M.; Cull, A.; Duez, N.J.; Filiberti, A.; Flechtner, H.; Fleishman, S.B.; de Haes, J.C.; et al. The European Organization for Research and Treatment of Cancer QLQ-C30: A quality-of-life instrument for use in international clinical trials in oncology. *J. Natl. Cancer Inst.* **1993**, *85*, 365–376. [CrossRef]
40. Cornette, T.; Vincent, F.; Mandigout, S.; Antonini, M.T.; Leobon, S.; Labrunie, A.; Venat, L.; Lavau-Denes, S.; Tubiana-Mathieu, N. Effects of home-based exercise training on VO2 in breast cancer patients under adjuvant or neoadjuvant chemotherapy (SAPA): A randomized controlled trial. *Eur. J. Phys. Rehabil. Med.* **2016**, *52*, 223–232.
41. Courneya, K.S.; Segal, R.J.; Mackey, J.R.; Gelmon, K.; Reid, R.D.; Friedenreich, C.M.; Ladha, A.B.; Proulx, C.; Vallance, J.K.; Lane, K.; et al. Effects of aerobic and resistance exercise in breast cancer patients receiving adjuvant chemotherapy: A multicenter randomized controlled trial. *J. Clin. Oncol. Off. J. Am. Soc. Clin. Oncol.* **2007**, *25*, 4396–4404. [CrossRef] [PubMed]
42. Husebo, A.M.; Dyrstad, S.M.; Mjaaland, I.; Soreide, J.A.; Bru, E. Effects of scheduled exercise on cancer-related fatigue in women with early breast cancer. *Sci. World J.* **2014**, *2014*, 271828. [CrossRef]
43. Mutrie, N.; Campbell, A.M.; Whyte, F.; McConnachie, A.; Emslie, C.; Lee, L.; Kearney, N.; Walker, A.; Ritchie, D. Benefits of supervised group exercise programme for women being treated for early stage breast cancer: Pragmatic randomised controlled trial. *BMJ* **2007**, *334*, 517. [CrossRef]
44. Travier, N.; Velthuis, M.J.; Steins Bisschop, C.N.; van den Buijs, B.; Monninkhof, E.M.; Backx, F.; Los, M.; Erdkamp, F.; Bloemendal, H.J.; Rodenhuis, C.; et al. Effects of an 18-week exercise programme started early during breast cancer treatment: A randomised controlled trial. *BMC Med.* **2015**, *13*, 121. [CrossRef] [PubMed]
45. Van Waart, H.; Stuiver, M.M.; van Harten, W.H.; Geleijn, E.; Kieffer, J.M.; Buffart, L.M.; de Maaker-Berkhof, M.; Boven, E.; Schrama, J.; Geenen, M.M.; et al. Effect of Low-Intensity Physical Activity and Moderate-to High-Intensity Physical Exercise During Adjuvant Chemotherapy on Physical Fitness, Fatigue, and Chemotherapy Completion Rates: Results of the PACES Randomized Clinical Trial. *J. Clin. Oncol. Off. J. Am. Soc. Clin. Oncol.* **2015**, *33*, 1918–1927. [CrossRef] [PubMed]
46. Van Vulpen, J.K.; Sweegers, M.G.; Peeters, P.H.M.; Courneya, K.S.; Newton, R.U.; Aaronson, N.K.; Jacobsen, P.B.; Galvao, D.A.; Chinapaw, M.J.; Steindorf, K.; et al. Moderators of Exercise Effects on Cancer-related Fatigue: A Meta-analysis of Individual Patient Data. *Med. Sci. Sports Exerc.* **2020**, *52*, 303–314. [CrossRef] [PubMed]
47. Furmaniak, A.C.; Menig, M.; Markes, M.H. Exercise for women receiving adjuvant therapy for breast cancer. *Cochrane Database Syst. Rev.* **2016**, *9*, CD005001. [CrossRef]
48. Courneya, K.S.; McKenzie, D.C.; Mackey, J.R.; Gelmon, K.; Friedenreich, C.M.; Yasui, Y.; Reid, R.D.; Cook, D.; Jespersen, D.; Proulx, C.; et al. Effects of exercise dose and type during breast cancer chemotherapy: Multicenter randomized trial. *J. Natl. Cancer Inst.* **2013**, *105*, 1821–1832. [CrossRef]
49. Sheehan, P.; Denieffe, S.; Murphy, N.M.; Harrison, M. Exercise is more effective than health education in reducing fatigue in fatigued cancer survivors. *Support. Care Cancer Off. J. Multinatl. Assoc. Support. Care Cancer* **2020**, *10*. [CrossRef]
50. An, K.Y.; Morielli, A.R.; Kang, D.W.; Friedenreich, C.M.; McKenzie, D.C.; Gelmon, K.; Mackey, J.R.; Reid, R.D.; Courneya, K.S. Effects of exercise dose and type during breast cancer chemotherapy on longer-term patient-reported outcomes and health-related fitness: A randomized controlled trial. *Int. J. Cancer* **2020**, *146*, 150–160. [CrossRef]
51. Morey, M.C.; Snyder, D.C.; Sloane, R.; Cohen, H.J.; Peterson, B.; Hartman, T.J.; Miller, P.; Mitchell, D.C.; Demark-Wahnefried, W. Effects of home-based diet and exercise on functional outcomes among older, overweight long-term cancer survivors: RENEW: A randomized controlled trial. *JAMA* **2009**, *301*, 1883–1891. [CrossRef] [PubMed]

52. Demark-Wahnefried, W.; Clipp, E.C.; Lipkus, I.M.; Lobach, D.; Snyder, D.C.; Sloane, R.; Peterson, B.; Macri, J.M.; Rock, C.L.; McBride, C.M.; et al. Main outcomes of the FRESH START trial: A sequentially tailored, diet and exercise mailed print intervention among breast and prostate cancer survivors. *J. Clin. Oncol. Off. J. Am. Soc. Clin. Oncol.* **2007**, *25*, 2709–2718. [CrossRef] [PubMed]
53. Demark-Wahnefried, W.; Clipp, E.C.; Morey, M.C.; Pieper, C.F.; Sloane, R.; Snyder, D.C.; Cohen, H.J. Lifestyle intervention development study to improve physical function in older adults with cancer: Outcomes from Project LEAD. *J. Clin. Oncol. Off. J. Am. Soc. Clin. Oncol.* **2006**, *24*, 3465–3473. [CrossRef]
54. Adams, R.N.; Mosher, C.E.; Blair, C.K.; Snyder, D.C.; Sloane, R.; Demark-Wahnefried, W. Cancer survivors' uptake and adherence in diet and exercise intervention trials: An integrative data analysis. *Cancer* **2015**, *121*, 77–83. [CrossRef]
55. Stone, C.R.; Friedenreich, C.M.; O'Reilly, R.; Farris, M.S.; Vallerand, J.R.; Kang, D.W.; Courneya, K.S. Predictors of Adherence to Different Volumes of Exercise in the Breast Cancer and Exercise Trial in Alberta. *Ann. Behav. Med. Publ. Soc. Behav. Med.* **2019**, *53*, 453–465. [CrossRef]
56. Buffart, L.M.; Sweegers, M.G.; May, A.M.; Chinapaw, M.J.; van Vulpen, J.K.; Newton, R.U.; Galvao, D.A.; Aaronson, N.K.; Stuiver, M.M.; Jacobsen, P.B.; et al. Targeting Exercise Interventions to Patients with Cancer in Need: An Individual Patient Data Meta-Analysis. *J. Natl. Cancer Inst.* **2018**, *110*, 1190–1200. [CrossRef]

nutrients

MDPI

Article

A Randomized Controlled Trial Testing the Effectiveness of Coping with Cancer in the Kitchen, a Nutrition Education Program for Cancer Survivors

Melissa Farmer Miller [1], Zhongyu Li [2] and Melissa Habedank [3,*]

[1] Department of Public Health, College of Health Sciences, Arcadia University, Glenside, PA 19038, USA; millerm@arcadia.edu

[2] Johns Hopkins Bloomberg School of Public Health, Baltimore, MD 21205, USA; lilizy597@gmail.com

[3] American Institute for Cancer Research, Arlington, VA 22209, USA

* Correspondence: m.habedank@aicr.org; Tel.: +1-800-843-8114

Received: 4 September 2020; Accepted: 12 October 2020; Published: 15 October 2020

Abstract: Following a diet rich in whole grains, vegetables, fruit, and beans may reduce cancer incidence and mortality. The aim of this study was to investigate the effect of Coping with Cancer in the Kitchen (CCK), an 8 week in-person program offering education, culinary demonstrations and food tasting, and psychosocial group support, compared to receiving CCK printed materials by mail on knowledge, confidence, and skills in implementing a plant-based diet. A total of 54 adult cancer survivors were randomly assigned to intervention (n = 26) and control groups (n = 27) with assessments at baseline, 9, and 15 weeks via self-administered survey. The response rate was 91% at 9 weeks and 58% at 15 weeks. The majority of our study participants were female breast cancer survivors (58%) who had overweight or obesity (65%). Compared with the control, there were significant ($p < 0.05$) increases in intervention participants' knowledge about a plant-based diet at weeks 9 and 15, reductions in perceived barriers to eating more fruits and vegetables at week 9, and enhanced confidence and skills in preparing a plant-based diet at week 15. There was a significant reduction in processed meat intake but changes in other food groups and psychosocial measures were modest. Participation in CCK in person increased knowledge, skills, and confidence and reduced barriers to adopting a plant-based diet. Positive trends in intake of plant-based foods and quality of life warrant further investigation in larger-scale studies and diverse populations.

Keywords: cancer survivors; health behavior intervention; diet and nutrition

1. Introduction

The overall aging of the United States population and changing prevalence of risk factors, including obesity, have increased the incidence of many types of cancer while advances in the early detection and treatment of cancer have led to reduced cancer mortality. These factors have combined to dramatically increase the number of cancer survivors [1,2]. Nearly 17 million people in the US were living with a history of a cancer diagnosis as of January 2019. This number is projected to grow to more than 22 million by 2030 and to more than 26 million by 2040 [1,3].

The importance of diet for cancer survivors is indicated by the accumulating evidence that a healthier diet after a cancer diagnosis can lead to improved treatment response, recovery, side-effect management, and disease outcomes [4–8]. The World Cancer Research Fund/American Institute for Cancer Research's (AICR's) Third Expert Report, *Diet, Nutrition, Physical Activity and Cancer: a Global Perspective*, states ten Cancer Prevention Recommendations [9]. Six of these recommendations focus on aspects of diet, including following a dietary pattern rich in whole grains, vegetables, fruit and beans, and limiting consumption of red meat and processed food, to reduce cancer incidence and mortality.

A substantial body of research has demonstrated the benefits of adherence to the AICR's Cancer Prevention Recommendations, including a lower risk of cancer incidence, recurrence, and death [10–14].

Despite the growing evidence supporting positive changes in diet to prevent cancer-related morbidity and mortality, most cancer survivors' adherence to the AICR's dietary recommendations is low. A recent systematic review and meta-analysis concluded that only 34% of cancer survivors met the recommendations for fruit and vegetable (F&V) intake and only 31% and 47% for fiber and red meat intake, respectively, while 87% of cancer survivors met the recommendations for smoking and 83% for alcohol intake [15], which suggests potential opportunities for interventions to improve the dietary patterns and decisions among cancer survivors. Diet/nutrition was recognized as a common concern in an observational survey research study of cancer patients and survivors in the community across the US [16] and among members of a community-based cancer care organization [17]. Further, cancer survivors frequently experience low quality of life partly attributed to nutritional problems and thus report high demand for nutritional support [18,19]. In fact, diagnoses of cancer often motivate individuals to make lifestyle changes [20–22]. Nonetheless, cancer survivors may find it difficult to act on their intentions and can struggle to achieve their goals when they are not given necessary tools such as specific action plans and evidence-based information for making and sustaining behavioral changes [20,23,24].

Coping with Cancer in the Kitchen (CCK) was initiated to help fill this gap between lifestyle recommendations and the sustained adoption of improved lifestyle behaviors among post-diagnosis and post-treatment cancer survivors. CCK is an 8 week in-person program offering multidisciplinary support for cancer survivors that includes nutrition education, culinary demonstrations and food tasting. It also offers facilitated group discussions with structured goal setting to address psychosocial health, and it introduces simple and effective techniques for coping with cancer-related stress. The first phase of research was conducted in 2017 to test the feasibility and acceptability and to determine the preliminary effect sizes of the CCK program. These promising results were used to inform refinements in both the CCK curriculum and the training [25]. In 2019, the next phase of research was conducted as a randomized controlled trial to rigorously evaluate the efficacy of the in-person CCK program versus receiving printed CCK materials. The objectives of this paper are to describe the randomized controlled trial, report the results, and examine whether the in-person CCK program (the intervention group) increased knowledge, skills, and confidence in adopting a plant-based diet as well as made positive changes in dietary intake and quality of life compared to the delivery of the CCK program through printed materials (the control group).

2. Materials and Methods

2.1. Program Overview

CCK was developed according to evidence-based concepts, including AICR's Cancer Prevention Recommendations, AICR's Foods that Fight Cancer™, and AICR's New American Plate®. CCK is a response to the needs of cancer survivors, registered dietitians, and other health professionals who lamented the lack of an evidence-based, standardized curriculum specifically for cancer survivors. A core principle guiding the design and development of CCK was to ensure that it benefitted communities in which it would be implemented. To achieve this goal, the design and development of CCK's evidence-based curriculum involved multiple partners across its first and second phases of research, including AICR, Living Plate (Far Hills, NJ, USA), Cancer Support Community (CSC; Washington, DC, USA), CSC of Central New Jersey (Bedminster, NJ, USA), registered dietitians and mental health practitioners. The CCK intervention was primarily guided by the Social Cognitive Theory (i.e., knowledge and skills development, self-efficacy, and observational learning mediate behavior change) [26] and the Transtheoretical Model (i.e., stages of change) [27,28]. Its psychosocial components were guided by CSC's affiliate model, which empowers individuals impacted by cancer to improve their health and well-being through active participation in community-based programs and active engagement with their health care team [29,30]. Furthermore, it is indicated that enhanced learning occurs in a small group defined by a shared cancer experience with professional and peer support to motivate behavior change [31,32].

Often the cost of food can be a barrier to trying new foods and purchasing perishable food items, both of which are important to adopting a plant-based diet. With this in mind, the CCK program's recipes were designed to include relatively basic, whole foods that can be found at standard grocery stores; the serving sizes are small to limit the quantity of ingredients needed and the potential for spoilage; and no name brands are explicitly recommended. As well, a variety of recipes at various price points are provided so that facilitators can choose to demonstrate the recipes that are most appropriate for their participants and communities, and the curriculum includes time for the facilitators and participants to discuss possible recipe variations and ingredient substitutions. Finally, each week, the facilitators probe for perceived barriers to adopting a plant-based diet related to each module topic (e.g., veggies, snacks, whole grains, and breakfasts) and addresses financial concerns.

The first CCK pilot study assessed the feasibility, acceptability, and pre–post-impact of the program by a single-arm intervention among 21 adult cancer survivors in 2017 [25]. Participants reported increased confidence preparing a variety of plant-based foods ($p = 0.002$), perceived control over cancer ($p = 0.034$), perception of dietary quality ($p = 0.009$), and weekly behavioral capability, including food and nutrition knowledge ($p < 0.001$). There was a non-significant (NS) trend towards increased F&V and whole grain intake with moderate effect sizes (0.2–0.5) for intake of beans and legumes, vegetables, and cooked whole grains like brown rice and quinoa. This single-arm pilot study achieved enrollment of 88% of the accrual target; program attendance at each session ranged from 48% to 100%. Participant satisfaction was positive with 100% of participants very satisfied (9 or 10 on 11-point Likert scale) with the cooking demonstrations, 93% very satisfied with the facilitated group discussions, and 87% very satisfied with the nutrition education.

2.2. Study Design

We conducted a two-arm, randomized controlled trial; cancer survivors were randomized to receive the 8-week CCK in-person multidisciplinary program immediately (intervention arm) or to receive CCK printed materials (control arm). Data was collected through self-administered patient surveys completed at baseline, post-intervention, and follow-up. A maximum total sample size of 60 (30 in each group) with 10% loss to follow-up and type I error of 5% provides >80% power to detect a difference (effect size) of at least 0.8 standard deviations between intervention and control groups (two-sample means test). This study was conducted according to the guidelines laid down in the Declaration of Helsinki, and all procedures involving research study participants were approved by Ethical and Independent Review Services (E&I) Institutional Review Board (Lee's Summit, MO, USA; Approval code: 19048-01). Written informed consent was obtained from all subjects/patients during enrollment. The trial was retrospectively registered with ClinicalTrials.gov (Identifier: NCT04528615).

2.3. Study Population

Cancer survivors were recruited from community members served by CSC-Los Angeles (LA, CA, USA) and Fanwood-Scotch Plains YMCA (Scotch Plains, NJ, USA) from April to June 2019, largely by social media, emailed letters of invitation to affiliate members, community-based presentations at cancer support groups, and fliers. The research sites in Los Angeles, CA, and Scotch Plains, NJ, were selected to represent a diversity of cancer survivors—from the west and east coasts of the US, respectively; from urban and small township/suburban regions, respectively; and from sites that provide programming specific to cancer or with a general community-based focus, respectively. Eligibility criteria included: (1) 18 years of age or older; (2) ever been diagnosed with cancer; (3) able to attend at least seven of the eight total sessions with mandatory first and last sessions; (4) willingness to be randomized and adhere to study protocol; (5) completed active cancer treatment (not including hormonal or other similar agents, e.g., tamoxifen). If potential candidates had not completed active cancer treatment, research staff determined eligibility if side effects of the current treatment had not affected sense of taste causing difficulty eating healthy foods, like F&V or whole grains, and had not caused a level of fatigue that would impede ability to attend an 8-week program, shop for healthy foods, and prepare recipes.

2.4. Randomization

Consenting participants were randomly assigned to 1 of 2 treatment groups: CCK intervention or printed materials control. A unique computer-generated list for each of the two research sites randomly sequenced intervention and control assignments. Allocation to the intervention or control groups was concealed until all participants had been enrolled at a site. Once the final participant was enrolled, research staff broke the treatment code and assigned participants in order of enrollment to either the intervention group or the control group by adding them to the randomized list of intervention and control assignments. The local research staff notified participants in the intervention group of the start date for the CCK program. Participants in the control group were notified of their group assignment and mailed a baseline survey with a postage paid return envelope approximately one week before the start of the in-person CCK program with instructions to complete the survey within two weeks.

2.5. CCK Onsite Teams

CCK was facilitated by a multidisciplinary team. Registered dietitians were trained to educate participants about AICR's Cancer Prevention Recommendations and two evidence-based programs: AICR's New American Plate® and AICR's Foods That Fight Cancer™. Licensed social workers were trained to facilitate group support and discussions that strategized how participants could overcome psychosocial barriers to nutrition behavior change in the context of cancer survivorship, equip them with strategies to help with cancer-related stress, and encourage goal setting. In addition, the registered dietitian or a culinary assistant demonstrated convenient, easy, and tasty ways to prepare and cook plant-based foods and offered tastings. Prior to facilitating CCK, facilitators completed a required two-part live virtual training about how to implement the CCK program with intervention fidelity, and they met weekly by phone with researchers using a semi-structured moderator guide to address any questions or problems that arose during program implementation. An evaluation of the training program indicated a high level of knowledge about CCK research procedures and preparedness to deliver program content.

2.6. CCK Intervention Group

Components of the CCK intervention are summarized in Table 1. CCK participants attended eight, in-person, 90-min classes convened weekly at their community-based organizational facility. The schedule at each site included a 1-week midterm break for a national holiday and staff travel, so the complete program extended to 9 weeks. Weekly themes included beans and whole grains, one-pot meals, breakfast and snacks, comfort foods, veggies, and building a Foods that Fight Cancer kitchen. To encourage attendance, some participants were sent email reminders before some classes and some absentees were contacted by phone.

Table 1. Description of intervention components and outcome measures used in a randomized controlled trial of Coping with Cancer in the Kitchen, a Nutrition Education Program for Cancer Survivors

Measure/Component	Description
	Intervention Components
Eight, in-person, 90-min group meetings convened weekly at community-based organizational facilities	
(1) Nutrition Education	Registered dietitians educated participants using slide presentations about the American Institute for Cancer Research's (AICR's) Recommendations for Cancer Prevention, "New American Plate®", and "Foods That Fight Cancer (FTFC)™" using 8 modules. Module 1: AICR Recommendations for Cancer Prevention Module 2: AICR's New American Plate and One-Pot Meals Module 3: Beans and Whole Grains Module 4: Breakfast and Snacks Module 5: Comfort Foods Module 6: Veggies Module 7: Building a FTFC Kitchen Module 8: Sharing and Caring Potluck
(2) Structured Group Learning and Support	Licensed social workers provided support through a structured and empowering group learning environment to address the complex, important (and, unfortunately, often rarely openly discussed) psychosocial barriers to nutrition behavior change in the context of cancer prevention, treatment, and survivorship.
(3) Cooking Demonstration	Culinary experts demonstrated convenient, easy, and tasty ways to prepare and cook FTFC and invited recipe tasting.
Sharing and Caring Potluck	The last in-person meeting of the program included a time to review the overall experience and engage in discussion. It was intended to explore milestones achieved, recognize precipitous moments of comprehension, connect to feelings related to the program ending, identify ongoing obstacles and/or challenges, identify changes and successes along the way, share ideas and hopes for continued success, and discuss take-aways from the group experience.
Recipe Cards	Each week participants received 2–3 printed recipe cards for foods exhibited and tasted during the culinary demonstration. Examples of recipes included Quinoa Salad, Everyday Green Smoothie, Southwestern Bean Salad, Buckwheat Cocoa-Chip Overnight Oats, Chili, and Whole Wheat Greek Pasta Salad.
Workbook	Pocket folders included written materials about dietary choices and recipe cards.
S.M.A.R.T Goal-Setting Worksheets	Each week participants completed a one-page worksheet that prompted them to identify one to three specific, measurable, actionable, relevant and time-bound goals that were revisited at the next session.
	Control Group Components
Printed Educational Materials	Participants in the control group received seven comprehensive summaries from Coping with Cancer in the Kitchen weekly module content and 14 recipe cards (two from each of the 7 weeks of culinary demonstrations). These were mailed to participants in one package upon completion of the baseline survey.
	Pre–Post Outcome Measures
Knowledge about a Plant-Based Diet	Participants rated their agreement (1 = *Strongly disagree*; 5 = *Strongly agree*) with six custom items developed by the research team, e.g., "I understand the benefits of consuming whole grains versus processed grains". A composite score was calculated as the average of the 6 ratings (range 1–5; Cronbach's alpha = 0.80).
Confidence Preparing a Variety of Plant Foods	Participants indicated "How sure are you that you could prepare the foods listed below in a tasty way?" (1 = *Very unsure*; 5 = *Very sure*). The 14-item scale included 4 whole grains; 4 beans, seeds and legumes; 3 green leafy vegetables; and 3 mixed foods, e.g., healthy one-pot meals. A composite score was calculated as the average of the 14 items (range 1–5; Cronbach's alpha = 0.75).

Table 1. *Cont.*

Measure/Component	Description
Skills to Practice a Plant-Based Diet	Participants rated their agreement (1 = *Strongly disagree*; 5 = *Strongly agree*) with five custom items developed by the research team, e.g., "I am confident that I can create a kitchen environment that makes it easier to store, prepare, and consume fruits, vegetables, whole grains, and beans."; the average of the five ratings was calculated to create a skills composite score (range 1–5; Cronbach's alpha = 0.88).
Barriers to Eating More Fruits and Vegetables and Whole Grains	We adapted items from an existing barriers instrument [33,34] to measure perceived barriers to eating more fruits and vegetables (F&V) (average score of 15 items; Cronbach's alpha = 0.89) and whole grains (average score of 14 items; Cronbach's alpha = 0.83). Participants were asked the general question, "Listed below are some common reasons why people don't eat more servings of *vegetables and fruits* each day. Indicate whether or not this is a reason for you by marking how much you agree or disagree." (1 = *Strongly disagree*; 5 = *Strongly agree*). In addition, using the same list of possible reasons (excluding *spoil too quickly*), participants indicated whether it was a common reason they did not eat more servings of *whole grains*. Example reasons included *take too much time to prepare; my family doesn't like them; hard to find a variety of good ones*.
Dietary Intake [Dietary Screener Questionnaire (DSQ) in the NHANES 2009-10] https: //epi.grants.cancer.gov/nhanes/dietscreen/	A 26-item dietary screener developed by the National Cancer Institute [35], which we shortened to include 17 questions that ask about the frequency of intake in the past month of F&V, whole grains, and processed and red meats. Scoring algorithms convert screener responses to estimates of daily intake of cup equivalents of F&V, including legumes and excluding French fries, and whole grains (ounce equivalents). Frequency responses to the processed meat question is converted to times per day. The DSQ provides a less burdensome alternative to 24-h recall when interest is in a limited set of dietary factors. We piloted use of the DSQ during the 2017 single-arm pilot study; intake was comparable to those participating in the National Health and Nutrition Examination Survey 2009–2010 [35].
General Quality of Life [a rapid version of the Functional Assessment of Cancer Therapy-General (FACT-G7)]	The Functional Assessment of Cancer Therapy-General (FACT-G) questionnaire is a general quality of life instrument that can be used to assess top-rated symptoms and concerns in cancer patients [36]. The FACT-G7 is a brief 7-item adaptation [37]. Internal consistency and reliability in the present study was good (Cronbach's alpha = 0.80).
Psychological Distress [4-item Patient Health Questionnaire for Depression and Anxiety (PHQ-4)]	The PHQ-4 is a brief 4-item validated screening scale for measuring core symptoms and signs of depression and anxiety [38].
Fatigue	Participants were asked to rate their level of fatigue on the average in the last week [0 = *Not at all fatigued*; 10 = *Fatigued as I could be*]. This item comes from the Fatigue Symptom Inventory that assesses the frequency and severity of fatigue and its perceived interference [39].
Emotional Support [NIH Toolbox® Emotional Support Fixed Form Age 18+ v 2.0, Short Form (SOC8)]	Participants completed the SOC8 which measures emotional support, or the perceived availability of someone to provide empathy or advice in times of need [40]. Higher scores represent more emotional support. Scores are converted to standardized *T* scores (mean = 50, standard deviation = 10); normative reference groups are the US general population.
Perceived Control over Course of Cancer	Participants were asked, "To what extent do you feel you have control over the course of your cancer (that is, whether your cancer will come back, get worse, or you will develop a different type of cancer)?" (0 = *No control at all*; 4 = *Complete control*).

2.7. Printed Materials Control Group

The control group received printed CCK educational materials including 7 written summaries of weekly nutrition content and 14 recipes that emphasized the weekly nutrition themes (see Table 1). (Only 7 written summaries were provided because the eighth session was a reflection and review session with no new nutrition information introduced.) Research staff mailed the materials to control participants upon completion of the baseline survey. The control group was not contacted again except for follow-up surveys at 9 and 15 weeks. Upon completion of the 9-week survey, control participants were emailed a $10 gift card.

2.8. Data Collection/Participant Survey

All data was collected through self-administered participant survey at baseline (pre-test, 0 week), post-intervention (9 week), and at follow-up (15 week). For participants in the CCK intervention group, the baseline and post-intervention surveys were completed in person at the beginning of the first and last CCK classes, respectively. The baseline and post-intervention surveys (with a postage-paid return envelope) were mailed to the control participants at their home within approximately one week of in-person CCK program commencement and conclusion. The 15-week online follow-up survey was completed by participants in both the intervention and control groups. A link to the online survey was emailed 6 weeks from the last session, or 15 weeks from baseline, with a window to complete the follow-up survey of 15 to 18 weeks. Participants were contacted by email or phone as a reminder to complete the questionnaires, as needed.

2.9. Baseline and Outcome Measures

The primary outcomes included knowledge about a plant-based diet (average score of 6 items; Cronbach's alpha = 0.80), confidence preparing a variety of plant-based foods (average score of 14 items; Cronbach's alpha = 0.75), and skills to practice a plant-based diet (average score of 5 items; Cronbach's alpha = 0.88) (see Table 1).

Secondary outcomes were measured using validated instruments for dietary intake (National Cancer Institute Dietary Screener Questionnaire) [35]; general quality of life (FACT-G7) [37]; psychological distress (PHQ-4) [38]; fatigue (single item from the Fatigue Symptom Inventory [39]); and emotional support (SOC8) [40]. We modified an existing scale [33,34] to measure perceived barriers to eating more F&V (average score of 15 items; Cronbach's alpha = 0.89) and whole grains (average score of 14 items; Cronbach's alpha = 0.83). 'Perceived barriers' was originally conceptualized as a moderating variable (those with fewer perceived barriers might experience a greater benefit from the CCK intervention) rather than an outcome variable and, for that reason, was not measured at 15 weeks follow-up. However, the findings indicate a reduction in barriers in the CCK intervention group, so we included 'perceived barriers' as an outcome.

Additional survey items included sociodemographic characteristics, disease characteristics, and health status.

2.10. Statistical Analysis

Descriptive statistics were calculated overall and by study group. Means and standard deviations are presented for continuous variables and frequencies and percentages are presented for categorical variables. We assessed the comparability between study groups using two-sample *t*-tests for continuous variables and Fisher's exact test for categorical variables and pre–post differences within study groups using paired *t*-tests. We used multiple regression analysis to estimate the difference between the CCK intervention and control groups at 9 weeks (post-intervention) and at 15 weeks (follow-up) adjusting for baseline (pre-test) levels of the dependent variable and research site (stratification variable). We considered a p-value < 0.05 statistically significant.

Effect-size calculations were also used as a standard for determining a meaningful treatment effect using Cohen's criteria for small, medium, and large effect sizes of 0.2, 0.5, and 0.8, respectively [41]. Standardizing the observed changes by the standard deviation (SD) allows for the comparison of the effect size magnitude across outcomes and can provide a meaningful reference for the future evaluation of the program in its implementation and dissemination. We calculated the *ES* statistic for the effect size, a form of Cohen's effect size index, as the mean of the changes in outcome scores for each study group at baseline and post-intervention (9 week) divided by the baseline SD [42]. Thus,

$$ES_{group} = \frac{\overline{x}_{Time2} - \overline{x}_{Time1}}{SD_{Time1}} \quad (1)$$

3. Results

The Consolidated Standards of Reporting Trials (CONSORT) flowchart for the trial is shown in Figure 1. A total of 54 adult cancer survivors were randomly assigned to intervention (n = 27) and control groups (n = 27). The majority (76%) of study participants learned about CCK from CSC-Los Angeles and YMCA staff; 9% from their oncologist; and 4% from other care providers. There was only 1 drop-out in the intervention arm (4%) who declined participation after randomization but before the CCK program commenced, indicating they could no longer commit to the duration of the program. The retention rate was 91% at 9 weeks and 58% at 15 weeks. There were no statistically significant differences between those who completed the online 15-week follow-up survey and those who did not with respect to sociodemographic variables, disease characteristics, and baseline levels of primary and secondary dependent variables. One participant in the intervention arm was excluded from the analysis due to an ineligibility discovered after completion of the program. Attendance rates ranged from 100% for the first session to 84% for sessions 3 and 8.

Characteristics for the total sample and by treatment group are shown in Table 2. Study participants were, on average, 61 years of age and primarily female breast cancer survivors with a college degree. The sample was 77% non-Hispanic white and 23% Hispanic or non-white race. Approximately half of the participants were married or living as married; 52% resided in suburban regions and 40% in urban areas. Most study subjects had overweight or obesity (65%), and less than half (45%) indicated they ate enough plant-based foods like fruits, vegetables, whole grains and beans in the past month ("most of the time" or "all of the time"). The CCK intervention group and printed materials control group were similar with respect to sociodemographic and disease characteristics.

Despite the randomization of participants, there were notable imbalances, though not statistically significant mean differences, between study groups in baseline levels of primary and secondary outcome measures (Table 3). Participants in the control group, on average, entered the study with higher confidence preparing a variety of plant-based foods and skills to practice a plant-based diet, and they also reported better quality of life, lower psychological distress, and less fatigue. Thus, we adjusted for baseline levels in regression analyses.

Knowledge about a plant-based diet significantly increased in the intervention arm (in-person CCK program) compared to the control arm (printed materials); this increase was sustained at 15 weeks post-intervention (Figure 2a). Confidence in preparing plant-based foods significantly increased at 15 weeks (Figure 2b) as did level of skills to practice a plant-based diet (Figure 2c). Perceived barriers to eating F&V decreased in the CCK intervention group and increased in the control group, and the adjusted difference between intervention and control groups was statistically significant (−0.37; 95% confidence interval (CI) −0.64, −0.10; see Table 3). In addition, there was a larger decrease in perceived barriers to consuming whole grains in the intervention group compared to the control group (−0.3 v −0.1), but the treatment effect did not reach statistical significance (−0.24; 95% CI −0.56, 0.07).

Participants began the study consuming, on average, 2.78 and 2.64 cup equivalents (5.6 and 5.3 servings) for F&V per day in the intervention and control groups, respectively. Intake of whole grains was approximately 1.3 and 1.4 servings per day, respectively. The between-arm differences in intake of F&V or whole grains consumption were not statistically significant. Further, intake increased in both groups over time with adjusted differences between groups post-intervention of 0.17 cup equivalents (95% CI −0.13, 0.47), or 0.34 servings, per day for F&V among participants in the intervention group compared with those in the control group, and 0.06 ounce equivalents (95% CI −0.18, 0.30), or 0.11 servings, in whole grains. The CCK intervention group also had significantly lower daily servings of processed meat in comparison to the control group at 9 and 15 weeks. In the intervention group, intake of processed meat was 0.14 times per day at baseline, which is equivalent to approximately one ($0.14 \times 7 = 0.98$) time per week, and it decreased to approximately one ($0.04 \times 30 = 1.2$) time in the past month.

No statistically significant differences between CCK intervention and control groups were observed in self-reported assessments of quality of life. Nonetheless, the baseline to 9-week change trended in a positive direction for general quality of life (+1.0 v +0.2; FACT-G7), psychological distress (−0.4 v +0.5; PHQ-4), and fatigue (−2.1 v −0.8; 11-point Likert). Similarly, the level of social support and perceived control over the course of cancer were relatively stable in both arms, and the results suggested a trend for increase at 15 weeks in the CCK intervention group.

The magnitude of effect sizes for the changes between baseline and 9 weeks (post-intervention) in the CCK intervention group, as measured by the *ES* statistic, are graphically presented in Figure 3. The effect sizes varied across categories of outcomes with large or nearly large effect sizes for outcomes measuring knowledge, confidence and skills. Barriers to consuming F&V and whole grains showed medium reductions. The effect sizes were in the range of small to medium for total F&V, whole grain and processed meat intake but were greater (with medium to large changes) for specific components of those dietary factors including beans and legumes, whole grain bread, and cooked whole grains (like quinoa). We detected small changes in quality of life measures. The decrease in fatigue was nonetheless large. We also considered using Cohen's *d* as a measure of effect size, which is the 9-week difference between the CCK intervention and control groups divided by the pooled standard deviation, which is common when comparing two independent groups. However, baseline differences between the CCK intervention and control groups were large relative to the variability, or standard deviation, of the factor of interest, and, therefore, may have underestimated the true treatment effect.

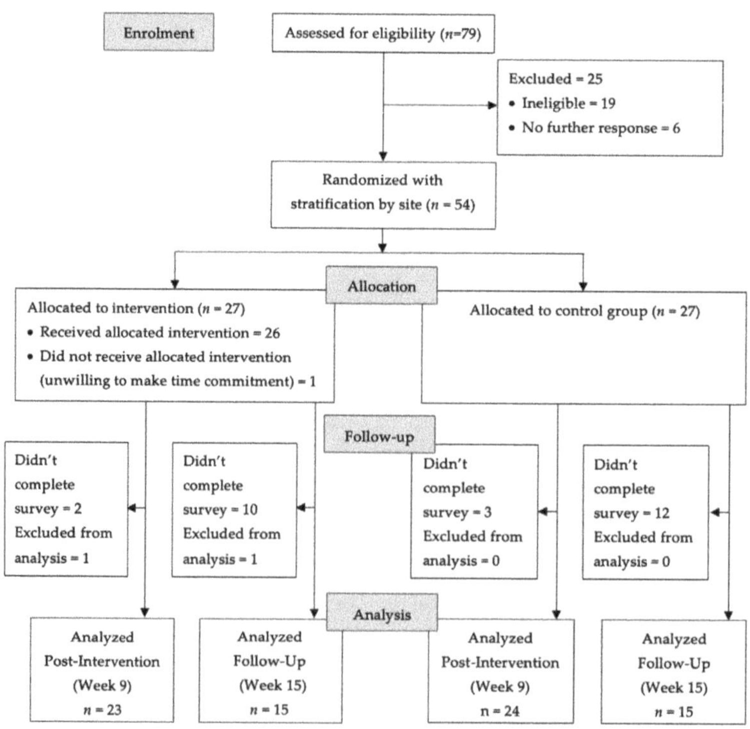

Figure 1. Coping with Cancer in the Kitchen trial flow diagram.

Table 2. Sociodemographic and disease characteristics of cancer survivors participating in a randomized study of Coping with Cancer in the Kitchen

	Total Sample (N = 53)		Coping with Cancer in the Kitchen Intervention (*n* = 26)		Printed Materials Control (*n* = 27)		
Factor	**No. of Participants**	**%**	**No. of Participants**	**%**	**No. of Participants**	**%**	***p*-Value [a]**
Sociodemographic Characteristics							
Age, years							0.25
Mean	61.2		59.5		62.8		
SD	10.5		9.7		11.1		
Range	37–80		37–74		41–80		
Gender							0.99
Female	49	92	24	92	25	93	
Male	4	8	2	8	2	7	
Race							0.43
Non-Hispanic white	41	77	18	69	23	85	
Non-Hispanic black	4	8	2	8	2	7	
Non-Hispanic, other	6	11	4	15	2	7	
Hispanic or Latino	2	4	2	8	0	0	
Education							0.30
Did not graduate college	14	26	6	23	8	30	
College graduate	16	30	6	23	10	37	
Postgraduate	23	43	14	54	9	33	
Geographic region							0.53
Urban	21	40	11	44	10	37	
Suburban	27	52	11	44	16	59	
Rural	3	6	2	8	1	4	
Marital status							0.43
Married or living as married	27	52	15	60	12	44	
Single (never married)	13	25	6	24	7	26	
Divorced or separated	9	17	4	16	5	19	
Widowed	3	6	0	0	3	11	
Disease Characteristics							
Time since cancer diagnosis							0.57
Mean	4.9		5.4		4.5		
SD	5.9		6.8		5.0		
Range	<1 to 27		<1 to 27		<1 to 21		

Table 2. *Cont.*

Factor	Total Sample (N = 53)		Coping with Cancer in the Kitchen Intervention (n = 26)		Printed Materials Control (n = 27)		p-Value [a]
	No. of Participants	%	No. of Participants	%	No. of Participants	%	
Primary cancer diagnosis							0.18
Breast	25	47	9	35	16	59	
Metastatic breast	6	11	5	19	1	4	
Blood	5	9	2	8	3	11	
Female reproductive	4	8	3	12	1	4	
Multiple cancers specified	6	11	2	8	4	15	
Other	7	13	5	19	2	7	
Ever received chemotherapy							
Yes	29	56	14	56	15	56	0.97
No	23	44	11	44	12	44	
Health Status							
Perceived health							0.87
Excellent	0	0	0	0	0	0	
Very good	15	31	8	33	7	28	
Good	23	47	11	46	12	48	
Fair	10	20	4	17	6	24	
Poor	1	2	1	4	0	0	
Body Mass Index, kg/m^2							0.84
Underweight (<18.5)	2	4	1	4	1	4	
Normal (18.5–24.9)	17	32	10	38	7	26	
Overweight (25.0–29.9)	13	25	6	23	7	26	
Obese (>30)	21	40	9	35	12	44	
In the last month, ate enough plant-based foods							0.73
None of the time	1	2	1	4	0	0	
A little of the time	7	14	4	17	3	12	
Some of the time	19	39	8	33	11	44	
Most of the time	16	33	9	38	7	28	
All of the time	6	12	2	8	4	16	

[a] Two-sample t-tests for continuous variables and Fisher's exact test for categorical variables. SD: standard deviation.

Table 3. Baseline, post-intervention (9-week), and follow-up (15-week) self-reported outcomes observed for the Coping with Cancer in the Kitchen intervention group and the printed materials control

Factor	Coping with Cancer in the Kitchen Intervention			Printed Materials Control			Adjusted Difference (95% CI) between Intervention and Control Groups [b]	
	Baseline ($n = 24$)	Post-Intervention ($n = 23$)	15-Week Follow-Up ($n = 15$)	Baseline ($n = 25$)	Post-Intervention ($n = 24$)	15-Week Follow-Up ($n = 15$)	Post-Intervention	15-Week Follow-Up
Primary Outcomes								
Knowledge about a Plant-Based Diet							**0.36 *** (0.06, 0.67)	**0.54 *** (0.11, 0.98)
Mean	3.6	4.4	4.2	3.7	4.1	3.9		
SD	0.5	0.5	0.5	0.8	0.5	0.6		
Within-arm mean difference		+0.8 [c,*]	+0.6 [d,*]		+0.4 [e]	+0.2 [f]		
Confidence Preparing a Variety of Plant Foods							0.36 (−0.02, 0.74)	**0.83 *** (0.23, 1.42)
Mean	3.4	4.0	4.1	3.7	3.8	3.7		
SD	0.9	0.8	1.0	1.0	0.8	0.7		
Within-arm mean difference		+0.6 *	+0.7 *		+0.1	0		
Skills to Practice a Plant-Based Diet							0.28 (−0.09, 0.64)	**0.65 *** (0.16, 1.14)
Mean	3.7	4.3	4.1	4.0	4.1	3.9		
SD	0.8	0.7	0.9	0.7	0.	0.7		
Within-arm mean difference		+0.6 *	+0.4 *		+0.1	−0.1		
Perceived Barriers								
Perceived Barriers to Eating More Fruits and Vegetables							**−0.37 *** (−0.64, −0.10)	NA
Mean	2.6	2.4	–	2.4	2.5	–		
SD	0.6	0.7	–	0.8	0.8	–		
Within-arm mean difference		−0.2 *	–		+0.1	–		
Perceived Barriers to Eating More Whole Grains							−0.24 (−0.56, 0.07)	NA

Table 3. *Cont.*

Factor	Coping with Cancer in the Kitchen Intervention			Printed Materials Control			Adjusted Difference (95% CI) between Intervention and Control Groups [b]	
	Baseline (n = 24)	Post-Intervention (n = 23)	15-Week Follow-Up (n = 15)	Baseline (n = 25)	Post-Intervention (n = 24)	15-Week Follow-Up (n = 15)	Post-Intervention	15-Week Follow-Up
Mean	2.6	2.3	–	2.6	2.5	–		
SD	0.5	0.7	–	0.6	0.7	–		
Within-arm mean difference		−0.3 *	–		−0.1	–		
Dietary Intake								
Total Fruit and Vegetable, cup equivalents per day [g]							0.17 (−0.13, 0.47)	−0.19 (−0.55, 0.17)
Mean	2.78	2.99	2.86	2.64	2.70	2.76		
SD	0.73	0.85	0.84	0.76	0.76	0.88		
Within-arm mean difference		+0.21	+0.08		+0.06	+0.12		
Whole Grains Total, ounce equivalents per day [h]							0.06 (−0.18, 0.30)	0.01 (−0.24, 0.27)
Mean	0.71	0.83	0.82	0.78	0.82	0.89		
SD	0.30	0.46	0.36	0.49	0.39	0.43		
Within-arm mean difference		+0.12	+0.11		+0.04	+0.11		
Processed Meat, times per day							**−0.08 *** (−0.15, −0.02)	**−0.09 *** (−0.18, −0.01)
Mean	0.14	0.04	0.04	0.08	0.10	0.11		
SD	0.23	0.05	0.05	0.15	0.18	0.17		
Range		−0.10 *	−0.10 *		+0.02	+0.03		
Quality of Life								
General Quality of Life, FACT-G7 Score							0.63 (−1.45, 2.71)	−0.09 (−3.54, 3.36)
Mean	17.1	18.1	17.1	17.8	18.0	18.4		
SD	5.1	5.0	6.1	5.9	5.7	4.2		
Within-arm mean difference		+1.0	0		+0.2	+0.6		

Table 3. *Cont.*

Factor	Coping with Cancer in the Kitchen Intervention			Printed Materials Control			Adjusted Difference (95% CI) between Intervention and Control Groups [b]	
	Baseline ($n = 24$)	Post-Intervention ($n = 23$)	15-Week Follow-Up ($n = 15$)	Baseline ($n = 25$)	Post-Intervention ($n = 24$)	15-Week Follow-Up ($n = 15$)	Post-Intervention	15-Week Follow-Up
Psychological Distress, PHQ-4							-0.83 (−2.13, 0.46)	−0.66 (−2.92, 1.60)
Mean	3.25	2.87	3.53	2.67	3.17	2.73		
SD	2.83	2.82	3.40	2.41	3.51	3.06		
Within-arm mean difference		−0.38	+0.28		+0.50	+0.06		
Fatigue, range 0–10							−0.76 (−2.09, 0.57)	−0.22 (−1.59, 1.14)
Mean	5.9	3.8	4.1	5.1	4.3	3.7		
SD	2.1	2.4	2.3	2.3	2.6	2.0		
Within-arm mean difference		−2.1 *	−1.8 *		−0.8	−1.4		
Emotional Support, SOC8 T-score							0.17 (−3.23, 3.57)	1.67 (−3.80, 7.13)
Mean	44.1	44.8	47.7	45.6	46.1	46.9		
SD	9.9	8.4	9.8	11.2	11.2	12.6		
Within-arm mean difference		+0.7	+3.6		+0.5	+1.3		
Perceived Control Over Course of Cancer, range 0–4							−0.18 (−0.63, 0.27)	0.34 (−0.20, 0.88)
Mean	1.4	1.5	1.9	1.1	1.4	1.0		
SD	1.1	0.9	0.8	1.2	1.0	1.1		
Within-arm mean difference		+0.1	+0.5		+0.3 *	−0.1		

* $p < 0.05$, paired t-test. * Bolded, $p < 0.05$, from linear regression analysis. [b] From linear regression analysis adjusting for baseline level and study site. Intent-to-treat analysis that used last observation carried forward for missing data did not appreciably change results or affect significance. [c] Difference computed from columns 3 and 2 (post-intervention minus baseline values) within the CCK group. [d] Difference computed from columns 4 and 2 (follow-up minus baseline values) within the CCK group. [e] Difference computed from columns 6 and 5 (post-intervention minus baseline values) within the control group. [f] Difference computed from columns 7 and 5 (follow-up minus baseline values) within the control group. [g] $\frac{1}{2}$ cup equivalent fruit and vegetable = 1 daily serving. [h] A total of ~0.56 oz equivalent whole grains = 1 daily serving. SD: standard deviation; CI: confidence interval; NA: not applicable; FACT-G7: a rapid version of the Functional Assessment of Cancer Therapy-General; PHQ-4: 4-item Patient Health Questionnaire for Depression and Anxiety; SOC8: NIH Toolbox® Emotional Support Fixed Form Age 18+ v 2.0, Short Form

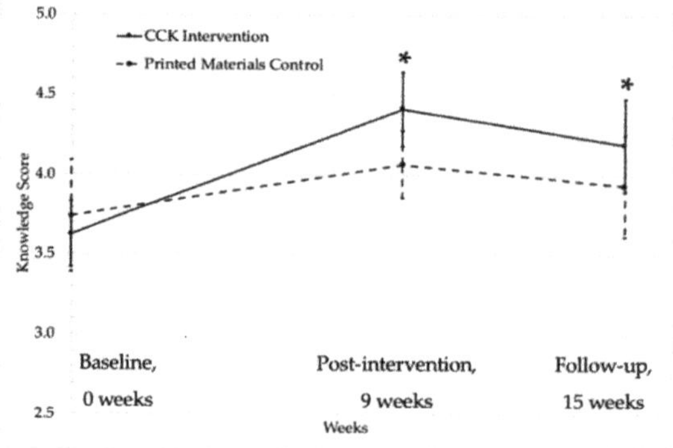

(a)

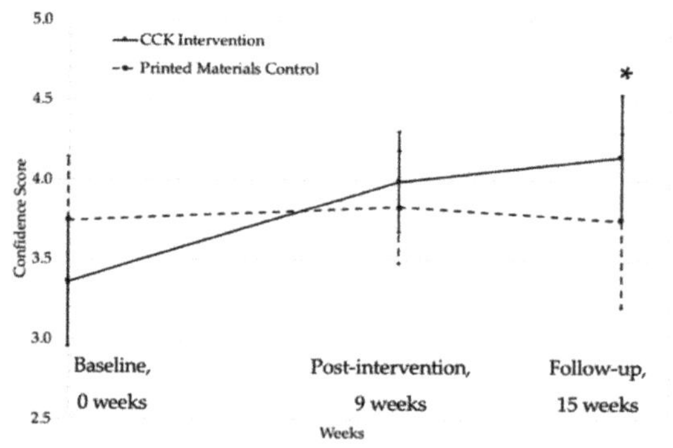

(b)

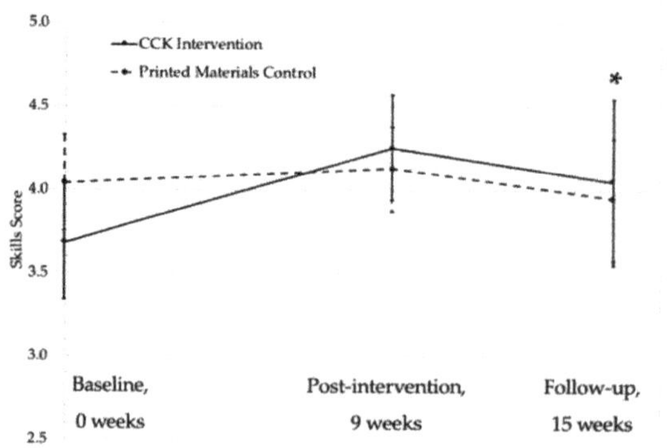

(c)

Figure 2. Change in (**a**) knowledge about a plant-based diet, (**b**) confidence preparing a variety of plant foods in a tasty way, and (**c**) skills to practice a plant-based diet for the Coping with Cancer in the Kitchen intervention group and the printed materials control group. * $p < 0.05$, difference between intervention and control groups adjusting for baseline level and study site. CCK: Coping with Cancer in the Kitchen.

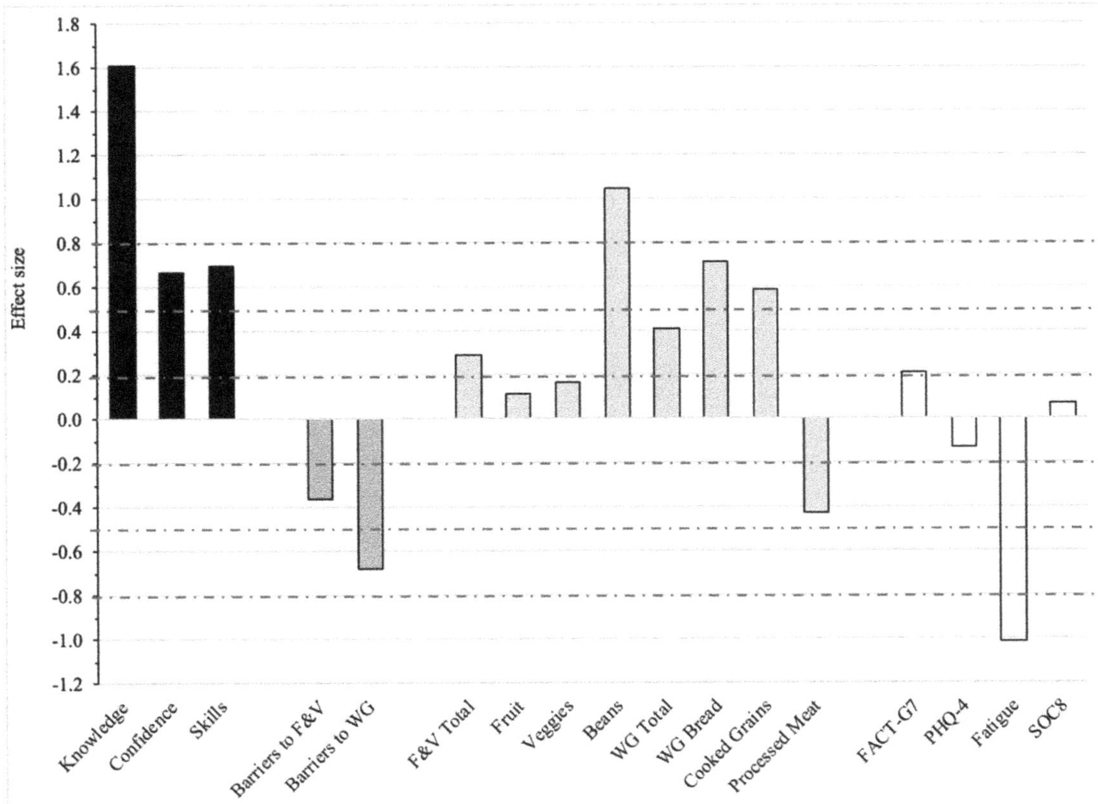

Figure 3. Effect size for the baseline to 9-week change in the Coping with Cancer in the Kitchen intervention group, as measured by the Effect Size statistic. The dotted lines correspond to small (0.2), medium (0.5), and large (0.8) effect sizes. WG: whole grain; F&V: fruit & vegetable; FACT-G7: a rapid version of the Functional Assessment of Cancer Therapy-General; PHQ-4: 4-item Patient Health Questionnaire for Depression and Anxiety; SOC8: NIH Toolbox® Emotional Support Fixed Form Age 18+ v 2.0, Short Form.

4. Discussion

This trial investigated the effectiveness of CCK, a multidisciplinary behavioral intervention incorporating both nutrition education and psychosocial support, in modulating several motivational, action, and environmental mediators for implementing a healthy plant-based diet and for improving quality of life among cancer survivors. Previously published interventions have shown that motivation, goal setting, action planning, social support, and instruction regarding how to perform desired behaviors are key elements in successfully promoting behavioral changes whereas self-monitoring is often less effective in doing so [43]. CCK reflects those concepts in its curriculum, and the results from this randomized controlled trial favored in-person delivery of CCK over receipt of CCK printed material only.

4.1. Primary Outcomes (Knowledge, Cooking Confidence, Skills) and Perceived Barriers

In-person delivery of the CCK program resulted in significant increases in knowledge, cooking confidence, and skills in adopting a plant-based diet over 9 and 15 weeks compared to the control group that received written CCK materials. Participants who attended CCK in person also reported a greater reduction in perceived barriers to the consumption of F&V and whole grains compared to the control group. As the literature indicates, people with higher perceived barriers tend to have poorer diets and are less likely to engage in behavioral changes, even when they are aware of the benefits of lifestyle changes [20,43,44]. Lack of access to accurate nutrition information, disbelief in diets and their relationship to cancer outcomes, low reinforcement from friends and family, and unfamiliarity with certain plant-based foods are commonly cited reasons for people not taking actions [20,44]. In particular, limited knowledge and skills in selecting and cooking healthy foods often demotivate cancer survivors from making dietary or lifestyle changes [44,45]. CCK addresses those concerns and obstacles by providing evidence-based nutrition education tailored to cancer survivors and delivered by registered dietitians. Moreover, CCK sessions included facilitator-led group discussions to enquire and consider approaches to reduce barriers to preparing and consuming a plant-based diet specifically for cancer survivors, and they contained weekly thematic cooking demonstrations using evidence-based AICR's dietary recommendations and inviting participants for recipe tastings.

Facilitated group discussion, access to trained facilitators, and experiential culinary support may contribute substantially to the observed difference between the intervention and control arms. Participants in the CCK intervention group were able to observe thematic, plant-based foods and recipes being prepared, ask questions, receive verbal information in real time about the health benefits of the ingredients in CCK recipes, interact with group members during the recipe demonstrations, and were encouraged to taste new foods. Fredericks et al. state that nutritional education with experiential features provides further drivers for behavioral change including collaboration, peer support, and palate development [46]. Though telephone and web/app-based interventions can be more accessible to a wider audience, especially in remote areas, they rely heavily on self-monitoring and often face challenges in retaining participants [47]. Conversely, CCK's in-person classes achieved high attendance (≥84%), which is comparable to other effective nutrition education programs targeting cancer survivors, such as Cocinar Para Tu Salud, a 12-week nutrition education program, and the Home Vegetable Gardening Interventions [48–50]. The study showed the benefit of in-person implementation over provision of printed materials only. However, given the current COVID-19 pandemic, future research could investigate virtual implementation of CCK using an online platform when in-person gatherings are prohibited, not possible or not preferred. The CCK program is available to survivors of all types of cancer. Its broad relevance increases the efficiency of delivery, the adaptability to local communities, and the scalability regionally and nationally.

The demonstrated enhancements in knowledge, skills, and confidence in practicing a plant-based diet at the end of the program, which continued their upward trend even at 15 weeks follow-up, implied, though not directly measured, the CCK effect on improving self-efficacy, which likely led to a higher level of patient empowerment [51,52]. Self-efficacy is a critical indicator of patient empowerment and a key construct of the Social Cognitive Theory and Transtheoretical Model used to guide the design and implementation of the CCK program for health behavior change [26,28,52]. Low self-efficacy (i.e., low confidence in one's ability to execute a course of action) is an important barrier impeding behavioral change among cancer survivors [53,54]. Further, practicing and experiencing are among the most important sources of self-efficacy [26]. We indeed observed that CCK's positive effect on self-efficacy was larger in the 15th week follow-up survey than immediately post-intervention (9 weeks) as participants had had more time to practice a plant-based diet by the 15th week. Higher self-efficacy has been indicated in studies to associate with higher probability of achieving and maintaining healthy behavioral goals and overall higher quality of life [55–57]. Further, empowerment also positively correlates with healthier behaviors and better decisions as well as health and clinical outcomes including improved disease management behaviors, use of health services, and health status [58,59].

4.2. Dietary Intake

The effects of CCK on total F&V and total whole grain intake were not statistically significant, but the observed net gain of 0.17 cup equivalents, or 0.34 servings, in daily F&V intake was similar to other studies of nutrition interventions designed to increase adult F&V intake [21,60]. A systematic review of the literature documented increases of 0.2 to 0.6 servings of F&V, and when targeting smaller focused communities, increases of 0.7 to 1.4 were observed [60]. The CCK participants had high baseline dietary intake of F&V at nearly 3 cups per day (exceeding the minimum intake recommended by AICR guidelines). Increasing people's intake of nutrients or foods when the baseline intake is already sufficient is expected to be challenging [21] and might have contributed to our study not observing significant post-interventional dietary changes. As hypothesized, we demonstrated a significant reduction in the consumption of processed meats in the CCK intervention group and observed medium to large effect sizes in specific components of total whole grain intake (e.g., cooked grains and bread). These changes, coupled with significant increases in mediators of behavior change (knowledge, confidence and skills) suggests that with longer follow-up, participants are likely to continue making important changes in adopting a plant-based diet. Additional investigation in diverse populations and communities whose adherence to the recommended dietary guidelines is low is warranted.

4.3. Quality of Life and Other Psychosocial Measures

There was modest impact on measurements of quality of life (QoL) in the current study, and effect sizes were generally small. Though not statistically significant, findings were within the range of effect sizes reported in prior research of interventions for cancer patients [61,62]. The data suggested positive trends in QoL, reduced fatigue and lower psychological distress at 9 weeks, though these trends did not reach statistical significance and were not sustained by the 15th week. Our limited sample size could be one of the major reasons for not finding statistical significance despite the observed positive trends. Likewise, a randomized controlled trial conducted by Uster and colleagues reported similar non-significant improvements in QoL among palliative cancer patients after nutrition and physical exercise interventions [63].

Living with cancer is undoubtedly stressful and associated with reduced QoL. Previous research demonstrated as many as 40% of cancer patients have clinically significant psychiatric comorbidities [64–66] and one in two reported significant distress [67]. Chronic stress is linked to several biobehavioral mechanisms related to the development of depressive symptoms and poorer cancer prognosis that may discourage people from making positive changes in their lifestyle; these factors may contribute to a vicious cycle of persistent emotional distress and accelerated physical deterioration [68]. Furthermore, it has been indicated that up to 60% of cancer survivors have never received psychosocial support due to limited access to such programs, suggesting the existence of unaddressed needs in this population [69]. We chose to measure health-related QoL using the FACT-G7 due to its brevity, validity, and reliability for use in cancer patients [37]. The Yanez (2013) study reported a mean score for the FACT-G7 of 18.0 among cancer patients and 19.5 in a general population sample [37]. In the current study, we reported an average baseline score of 17.1 among participants in the CCK intervention group with a one-point increase at 9 weeks post-intervention. The CCK intervention has the potential to close the gap between cancer survivors and the general population in health-related quality of life and mental health directly through learning coping strategies to reduce stress and indirectly through improving diet quality [70]. Furthermore, in a naturalistic study, Giese-Davis et al. showed that CSC therapist-led support groups provided an experience in which the development of a new attitude was valued [71].

4.4. Limitations

This trial was small, and a substantial proportion of participants did not complete the online follow-up survey at 15 weeks. The low rate of response to completing the 15-week survey may in part be attributed to the online platform, which was a departure from the pen-and-paper format of the previous surveys. Retention rates may have been higher if we had disseminated the questionnaire in the same format as the previous and additionally provided an incentive upon completion of the survey, either financial or educational by disseminating preliminary study findings to interested participants.

Limitations in our dietary measurement methods may also have prevented us from fully uncovering the true effects of CCK on dietary intake. The use of 24-h recall, food records, and objective biomarkers may be more sensitive to changes and warranted in future evaluations of CCK, which would allow us to quantify more nuanced changes in dietary intakes, such as replacing processed grains with whole grains rather than an overall increase in whole grains. Furthermore, follow-up was short. We recognize that behavioral change and changes in quality of life may take longer than 8–15 weeks. We hypothesized that participation in the CCK intervention would indirectly influence health-related behaviors through self-efficacy mediators. With more time to practice acquired skills, intake of plant-based foods may continue to increase in the CCK intervention arm with longer follow-up.

Another limitation of the study was that clinical outcomes were only measured through patient-report. We did not include body mass index or percent body fat as outcomes since the CCK program was not specifically designed to be a weight loss program. While adopting a healthier diet can result in weight loss or changes to body composition, the goals of cancer survivors can be varied with some aiming to increase body weight and others to lose or maintain weight. As the program is more widely disseminated and administered in academic and clinical settings, there may be increased feasibility and interest in measuring pro- and anti-inflammatory biomarkers (e.g., C-reactive protein, Interleukins 3, 6, 8, and 10, and Tumor Necrosis Factor-α) and plasma concentration of antioxidants.

Other limitations included self-selected samples of participants who are predominantly female, White, fairly educated, and a substantial proportion of individuals with breast cancer. These limitations impact the study results' generalizability to a more diverse population. However, while this sample is not representative of all cancer patients and survivors across the US [72], it is representative of those who tend to seek nutrition education as well as social and emotional support in their community. Future work is required to understand the impact that CCK has on participants' long-term behavior changes, and to evaluate its applicability and cultural sensitivity among other diverse samples and settings.

5. Conclusions

Participation in the in-person CCK intervention led to improvements in nutrition and food-related knowledge and skills as well as confidence in adopting a plant-based diet among cancer survivors. As an evidence-based, experiential nutrition education program that embeds psychosocial health elements, CCK can serve as the standard for a high-quality community-based survivorship program. Further studies and larger-scale implementation of CCK in more diverse populations are needed to further understand its effect on health-related behaviors and wellbeing. Promoting lifestyle behavioral change through programs like CCK has potential and value to improve cancer survivors' nutritional status and quality of life.

Author Contributions: Conceptualization, M.F.M. and M.H.; methodology, M.F.M. and M.H.; formal analysis, M.F.M.; investigation, M.F.M. and M.H.; resources, M.F.H.; data curation, M.F.M.; writing—original draft preparation, M.F.M. and Z.L.; writing—review and editing, M.F.M., Z.L., and M.H.; visualization, M.F.M. and Z.L.; supervision, M.F.M. and M.H.; project administration, M.H.; funding acquisition, M.H. All authors have read and agreed to the published version of the manuscript.

Funding: The American Institute for Cancer Research.

Acknowledgments: Executive Investigators: Alice Bender, MS, RDN, American Institute for Cancer Research; Jill Kaplan, LCSW, Cancer Support Community Central New Jersey; Deirdre McGinley-Gieser, American Institute for Cancer Research; Jeanne Petrucci, MS, RDN, Living Plate; Amy Sutton, Cancer Support Community Central New Jersey. Participating Investigators: Nicole Angel, Cancer Support Community Los Angeles; Christopher Anrig, LCSW, Independent Contractor; Sherilyn Cognetti, Fanwood-Scotch Plains YMCA; Shannon La Cava, PsyD, Cancer Support Community Los Angeles; Donna Peart, MS, RDN, Fanwood-Scotch Plains YMCA; Tamar Rothenberg, MS, RDN, Touro College. The authors thank Nigel Brockton, American Institute for Cancer Research, for his helpful advice on this manuscript.

Conflicts of Interest: The authors declare no conflict of interest. AICR provided funding for project costs but did not influence the analyses and was blinded to all results until after the completion of all analyses.

References

1. Bluethmann, S.M.; Mariotto, A.B.; Rowland, J.H. Anticipating the "Silver Tsunami": Prevalence Trajectories and Comorbidity Burden among Older Cancer Survivors in the United States. *Cancer Epidemiol. Biomark. Prev.* **2016**, *25*, 1029–1036. [CrossRef]
2. Gallagher, E.J.; LeRoith, D. Obesity and Diabetes: The Increased Risk of Cancer and Cancer-Related Mortality. *Physiol. Rev.* **2015**, *95*, 727–748. [CrossRef] [PubMed]
3. Miller, K.D.; Nogueira, L.; Mariotto, A.B.; Rowland, J.H.; Yabroff, K.R.; Alfano, C.M.; Jemal, A.; Kramer, J.L.; Siegel, R.L. Cancer Treatment and Survivorship Statistics, 2019. *CAA Cancer J. Clin.* **2019**, *69*, 363–385. [CrossRef] [PubMed]
4. van Zutphen, M.; Boshuizen, H.C.; Kok, D.E.; van Baar, H.; Geijsen, A.J.M.R.; Wesselink, E.; Winkels, R.M.; van Halteren, H.K.; de Wilt, J.H.W.; Kampman, E.; et al. Colorectal Cancer Survivors Only Marginally Change Their Overall Lifestyle in the First 2 Years Following Diagnosis. *J. Cancer Surviv.* **2019**. [CrossRef] [PubMed]
5. Ornish, D.; Magbanua, M.J.M.; Weidner, G.; Weinberg, V.; Kemp, C.; Green, C.; Mattie, M.D.; Marlin, R.; Simko, J.; Shinohara, K.; et al. Changes in Prostate Gene Expression in Men Undergoing an Intensive Nutrition and Lifestyle Intervention. *Proc. Natl. Acad. Sci. USA* **2008**, *105*, 8369–8374. [CrossRef]
6. Ornish, D.; Lin, J.; Daubenmier, J.; Weidner, G.; Epel, E.; Kemp, C.; Magbanua, M.J.M.; Marlin, R.; Yglecias, L.; Carroll, P.R.; et al. Increased Telomerase Activity and Comprehensive Lifestyle Changes: A Pilot Study. *Lancet Oncol.* **2008**, *9*, 1048–1057. [CrossRef]
7. Bodai, B.I.; Tuso, P. Breast Cancer Survivorship: A Comprehensive Review of Long-Term Medical Issues and Lifestyle Recommendations. *Perm. J.* **2015**, *19*, 48–79. [CrossRef]
8. Schwedhelm, C.; Boeing, H.; Hoffmann, G.; Aleksandrova, K.; Schwingshackl, L. Effect of Diet on Mortality and Cancer Recurrence among Cancer Survivors: A Systematic Review and Meta-Analysis of Cohort Studies. *Nutr. Rev.* **2016**, *74*, 737–748. [CrossRef]
9. World Cancer Research Fund/American Institute for Cancer Research. Diet, Nutrition, Physical Activity and Cancer: A Global Perspective. *Continuous Update Project Expert Report.* 2018. Available online: https://www.wcrf.org/dietandcancer (accessed on 4 September 2020).
10. Ferrini, K.; Ghelfi, F.; Mannucci, R.; Titta, L. Lifestyle, Nutrition and Breast Cancer: Facts and Presumptions for Consideration. *Ecancermedicalscience* **2015**, *9*, 1–11. [CrossRef]
11. Inoue-Choi, M.; Robien, K.; Lazovich, D. Adherence to the WCRF/AICR Guidelines for Cancer Prevention is Associated with Lower Mortality among Older Female Cancer Survivors. *Cancer Epidemiol. Biomark. Prev.* **2013**, *22*, 792–802. [CrossRef] [PubMed]
12. Jankovic, N.; Geelen, A.; Winkels, R.M.; Mwungura, B.; Fedirko, V.; Jenab, M.; Illner, A.K.; Brenner, H.; Ordonez-Mena, J.M.; De Jong, J.C.K.; et al. Adherence to the WCRF/AICR Dietary Recommendations for Cancer Prevention and Risk of Cancer in Elderly from Europe and the United States: A Meta-Analysis within the CHANCES Project. *Cancer Epidemiol. Biomark. Prev.* **2017**, *26*, 136–144. [CrossRef] [PubMed]

13. Solans, M.; Chan, D.S.M.; Mitrou, P.; Norat, T.; Romaguera, D. A Systematic Review and Meta-Analysis of the 2007 WCRF/AICR Score in Relation to Cancer-Related Health Outcomes. *Ann. Oncol.* **2020**, *31*, 352–368. [CrossRef] [PubMed]
14. Romaguera, D.; Ward, H.; Wark, P.A.; Vergnaud, A.C.; Peeters, P.H.; van Gils, C.H.; Ferrari, P.; Fedirko, V.; Jenab, M.; Boutron-Ruault, M.C.; et al. Pre-Diagnostic Concordance with the WCRF/AICR Guidelines and Survival in European Colorectal Cancer Patients: A Cohort Study. *BMC Med.* **2015**, *13*, 1–12. [CrossRef]
15. Tollosa, D.N.; Tavener, M.; Hure, A.; James, E.L. Adherence to Multiple Health Behaviours in Cancer Survivors: A Systematic Review and Meta-Analysis. *J. Cancer Surviv.* **2019**, *13*, 327–343. [CrossRef]
16. Buzaglo, J.S.; Zaleta, A.K.; McManus, S.; Golant, M.; Miller, M.F. Cancer Support Source (R): Validation of a Revised Multi-Dimensional Distress Screening Program for Cancer Patients and Survivors. *Supportive Care Cancer* **2019**. [CrossRef]
17. Miller, M.F.; Mullins, C.D.; Onukwugha, E.; Golant, M.; Buzaglo, J.S. Discriminatory Power of a 25-Item Distress Screening Tool: A Cross-Sectional Survey of 251 Cancer Survivors. *Qual. Life Res.* **2014**, *23*, 2855–2863. [CrossRef] [PubMed]
18. Koshimoto, S.; Arimoto, M.; Saitou, K.; Uchibori, M.; Hashizume, A.; Honda, A.; Amano, K.; Nakajima, Y.; Uetake, H.; Matsushima, E. Need and Demand for Nutritional Counselling and Their Association with Quality of Life, Nutritional Status and Eating-Related Distress among Patients with Cancer Receiving Outpatient Chemotherapy: A Cross-Sectional Study. *Supportive Care Cancer* **2019**, *27*, 3385–3394. [CrossRef]
19. Kotronoulas, G.; Papadopoulou, C.; Burns-Cunningham., K.; Simpson, M.; Maguire, R. A systematic review of the supportive care needs of people living with and beyond cancer of the colon and/or rectum. *Eur. J. Oncol. Nurs.* **2017**, *29*, 60–70. [CrossRef]
20. Beeken, R.J.; Williams, K.; Wardle, J.; Croker, H. "What about Diet?" A Qualitative Study of Cancer Survivors' Views on Diet and Cancer and Their Sources of Information. *Eur. J. Cancer Care Engl.* **2016**, *25*, 774–783. [CrossRef]
21. Campbell, M.K.; Carr, C.; Devellis, B.; Switzer, B.; Biddle, A.; Amamoo, M.A.; Walsh, J.; Zhou, B.; Sandler, R. A Randomized Trial of Tailoring and Motivational Interviewing to Promote Fruit and Vegetable Consumption for Cancer Prevention and Control. *Ann. Behav. Med.* **2009**, *38*, 71–85. [CrossRef]
22. Demark-Wahnefried, W.; Aziz, N.M.; Rowland, J.H.; Pinto, B.M. Riding the Crest of the Teachable Moment: Promoting Long-Term Health after the Diagnosis of Cancer. *J. Clin. Oncol.* **2005**, *23*, 5814–5830. [CrossRef] [PubMed]
23. Contento, I.R. Nutrition Education: Linking Research, Theory, and Practice. *Asia Pac. J. Clin. Nutr.* **2008**, *17*, 176–179. [PubMed]
24. Coa, K.I.; Smith, K.C.; Klassen, A.C.; Caulfield, L.E.; Helzlsouer, K.; Peairs, K.; Shockney, L. Capitalizing on the "Teachable Moment" to Promote Healthy Dietary Changes among Cancer Survivors: The Perspectives of Health Care Providers. *Supportive Care Cancer* **2015**, *23*, 679–686. [CrossRef] [PubMed]
25. Miller, M.F.; Bender, A.G.; Feeney, C.; Kaplan, J.; McGinley-Gieser, D.; Petrucci, J.; Santangelo, A.; Saxton, C.; Soult, B.; Sutton, A.; et al. Evaluation of a Community-Based Experiential Nutrition and Cooking Education Program for Cancer Survivors. *Ann. Behav. Med.* **2018**, *52*, S497.
26. Bandura, A. Self-Efficacy: Toward a Unifying Theory of Behavioral Change. *Psychol. Rev.* **1977**, *84*, 191–215. [CrossRef] [PubMed]
27. Prochaska, J.O.; Velicer, W.F. The Transtheoretical Model of Health Behavior Change. *Am. J. Health Promot.* **1997**, *12*, 38–48. [CrossRef]
28. Graffigna, G.; Barello, S. Spotlight on the Patient Health Engagement Model (PHE Model): A Psychosocial Theory to Understand People's Meaningful Engagement in Their Own Health Care. *Patient Prefer. Adherence* **2018**, *12*, 1261–1271. [CrossRef]
29. Golant, M.; Thiboldeaux, K. The Wellness Community's Integrative Model of Evidence-Based Psychosocial Programs, Services, and Interventions. *Psycho-Oncology* **2010**. [CrossRef]
30. Golant, M.; Zaleta, A.; Ash-Lee, S.; Buzaglo, J.; Stein, K.; Saxton, C.; Donzinger, M.; Thiboldeaux, K.; House, L. The Engaged Patient: The Cancer Support Community's Comprehensive Model of Psychosocial Programs, Services and Research. In *PsychoOncology*, 4th ed.; Breitbart, W.S., Butow, P.N., Jacobsen, P.B., Lam, W., Lazenby, M., Loscalzo, M.J., Eds.; Oxford University Press: Oxford, UK, 2021.
31. Leszcz, M.; Goodwin, P.J. The Rationale and Foundations of Group Psychotherapy for Women with Metastatic Breast Cancer. *Int. J. Group Psychother.* **1998**, *48*, 245–273. [CrossRef]

32. Borek, A.J.; Abraham, C. How Do Small Groups Promote Behaviour Change? An Integrative Conceptual Review of Explanatory Mechanisms. *Appl. Psychol. Health Well-Being* **2018**, *10*, 30–61. [CrossRef]
33. Demark-Wahnefried, W.; Clipp, E.C.; Mcbride, C.; Lobach, D.F.; Lipkus, I.; Peterson, B.; Clutter Snyder, D.; Sloane, R.; Arbanas, J.; Kraus, W.E. Design of FRESH START: A Randomized Trial of Exercise and Diet among Cancer Survivors. *Med. Sci. Sport. Exerc.* **2003**, *35*. [CrossRef] [PubMed]
34. Vijan, S.; Stuart, N.S.; Fitzgerald, J.T.; Ronis, D.L.; Hayward, R.A.; Slater, S.; Hofer, T.P. Barriers to Following Dietary Recommendations in Type 2 Diabetes. *Diabet. Med.* **2005**, *22*, 32–38. [CrossRef] [PubMed]
35. Thompson, F.E.; Midthune, D.; Kahle, L.; Dodd, K.W. Development and Evaluation of the National Cancer Institute's Dietary Screener Questionnaire Scoring Algorithms. *J. Nutr.* **2017**, *147*, 1226–1233. [CrossRef]
36. Cella, D.F.; Tulsky, D.S.; Gray, G.; Sarafian, B.; Linn, E.; Bonomi, A.; Silberman, M.; Yellen, S.B.; Winicour, P.; Brannon, J. The Functional Assessment of Cancer Therapy Scale: Development and Validation of the General Measure. *J. Clin. Oncol.* **1993**, *11*, 570–579. [CrossRef]
37. Yanez, B.; Pearman, T.; Lis, C.G.; Beaumont, J.L.; Cella, D. The FACT-G7: A Rapid Version of the Functional Assessment of Cancer Therapy-General (FACT-G) for Monitoring Symptoms and Concerns in Oncology Practice and Research. *Ann. Oncol.* **2013**, *24*, 1073–1078. [CrossRef]
38. Kroenke, K.; Spitzer, R.L.; Williams, J.B.W.; Löwe, B. An Ultra-Brief Screening Scale for Anxiety and Depression: The PHQ-4. *Psychosomatics* **2009**, *50*, 613–621. [CrossRef]
39. Hann, D.M.; Jacobsen, P.B.; Azzarello, L.M.; Martin, S.C.; Curran, S.L.; Fields, K.K.; Greenberg, H.; Lyman, G. Measurement of Fatigue in Cancer Patients: Development and Validation of the Fatigue Symptom Inventory. *Qual. Life Res.* **1998**, *7*, 301–310. [CrossRef] [PubMed]
40. Cyranowski, J.M.; Zill, N.; Bode, R.; Butt, Z.; Kelly, M.A.R.; Pilkonis, P.A.; Salsman, J.M.; Cella, D. Assessing Social Support, Companionship, and Distress: National Institute of Health (NIH) Toolbox Adult Social Relationship Scales. *Health Psychol.* **2013**, *32*, 293–301. [CrossRef]
41. Cohen, J. *Statistical Power Analysis for the Behavioral Sciences*; Lawrence Erlbaum Associates: Hillsdale, NJ, USA, 1988; ISBN 978-0-805-80283-2.
42. Machin, D.; Fayer, P.M. *Quality of Life: The Assessment, Analysis and Interpretation of Patient-Reported Outcomes*; John Wiley and Sons: Chichester, UK, 2013; ISBN 978-1-118-69945-4.
43. Greenlee, H.; Santiago-Torres, M.; McMillen, K.K.; Ueland, K.; Haase, A.M. Helping Patients Eat Better During and Beyond Cancer Treatment. *Cancer J.* **2019**, *25*, 320–328. [CrossRef]
44. Lee, M.K.; Park, S.Y.; Choi, G.S. Facilitators and Barriers to Adoption of a Healthy Diet in Survivors of Colorectal Cancer. *J. Nurs. Scholarsh.* **2019**, *51*, 509–517. [CrossRef]
45. Corbett, T.; Cheetham, T.; Müller, A.M.; Slodkowska-Barabasz, J.; Wilde, L.; Krusche, A.; Richardson, A.; Foster, C.; Watson, E.; Little, P.; et al. Exploring Cancer Survivors' Views of Health Behaviour Change: "Where Do You Start, Where Do You Stop with Everything?". *Psychooncology* **2018**, *27*, 1816–1824. [CrossRef] [PubMed]
46. Fredericks, L.; Koch, P.A.; Liu, A.; Galitzdorfer, L.; Costa, A.; Utter, J. Experiential Features of Culinary Nutrition Education That Drive Behavior Change: Frameworks for Research and Practice. *Health Promot. Pract.* **2020**, *704*, 1–5. [CrossRef] [PubMed]
47. Lynch, S.M.; Stricker, C.T.; Brown, J.C.; Berardi, J.M.; Vaughn, D.; Domchek, S.; Filseth, S.; Branas, A.; Weiss-Trainor, E.; Schmitz, K.H.; et al. Evaluation of a Web-Based Weight Loss Intervention in Overweight Cancer Survivors Aged 50 Years and Younger. *Obes. Sci. Pract.* **2017**, *3*, 83–94. [CrossRef]
48. Greenlee, H.; Gaffney, A.O.; Aycinena, A.C.; Koch, P.; Contento, I.; Karmally, W.; Richardson, J.M.; Shi, Z.; Lim, E.; Tsai, W.Y.; et al. Long-Term Diet and Biomarker Changes after a Short-Term Intervention among Hispanic Breast Cancer Survivors: The Cocinar Para Su Salud! Randomized Controlled Trial. *Cancer Epidemiol. Biomark. Prev.* **2016**, *25*, 1491–1502. [CrossRef] [PubMed]
49. Aycinena, A.C.; Jennings, K.A.; Gaffney, A.O.; Koch, P.A.; Contento, I.R.; Gonzalez, M.; Guidon, E.; Karmally, W.; Hershman, D.; Greenlee, H. Cocinar Para Su Salud! Development of a Culturally Based Nutrition Education Curriculum for Hispanic Breast Cancer Survivors Using a Theory-Driven Procedural Model. *Health Educ. Behav.* **2017**, *44*, 13–22. [CrossRef] [PubMed]

50. Demark-Wahnefried, W.; Cases, M.G.; Cantor, A.B.; Frugé, A.D.; Smith, K.P.; Locher, J.; Cohen, H.J.; Tsuruta, Y.; Daniel, M.; Kala, R.; et al. Pilot Randomized Controlled Trial of a Home Vegetable Gardening Intervention among Older Cancer Survivors Shows Feasibility, Satisfaction, and Promise in Improving Vegetable and Fruit Consumption, Reassurance of Worth, and the Trajectory of Central Adiposity. *J. Acad. Nutr. Diet.* **2018**, *118*, 689–704. [CrossRef]
51. Funnell, M.M.; Anderson, R.M.; Arnold, M.S.; Barr, P.A.; Donnelly, M.; Johnson, P.D.; Taylor-Moon, D.; White, N.H. Empowerment: An Idea Whose Time Has Come in Diabetes Education. *Diabetes Educ.* **1991**, *17*, 37–41. [CrossRef]
52. Bravo, P.; Edwards, A.; Barr, P.J.; Scholl, I.; Elwyn, G.; McAllister, M. Conceptualising Patient Empowerment: A Mixed Methods Study. *BMC Health Serv. Res.* **2015**, *15*. [CrossRef]
53. Basen-Engquist, K.; Alfano, C.M.; Maitin-Shepard, M.; Thomson, C.A.; Schmitz, K.H.; Pinto, B.M.; Stein, K.; Zucker, D.S.; Syrjala, K.L.; Fallon, E.; et al. Agenda for Translating Physical Activity, Nutrition, and Weight Management Interventions for Cancer Survivors into Clinical and Community Practice. *Obesity* **2017**, *25*, S9–S22. [CrossRef]
54. Chirico, A.; Lucidi, F.; Merluzzi, T.; Alivernini, F.; De Laurentiis, M.; Botti, G.; Giordano, A. A Meta-Analytic Review of the Relationship of Cancer Coping Selfefficacy with Distress and Quality of Life. *Oncotarget* **2017**, *8*, 36800–36811. [CrossRef]
55. Dockham, B.; Schafenacker, A.; Yoon, H.; Ronis, D.L.; Kershaw, T.; Titler, M.; Northouse, L. Implementation of a Psychoeducational Program for Cancer Survivors and Family Caregivers at a Cancer Support Community Affiliate: A Pilot Effectiveness Study. *Cancer Nurs.* **2016**, *39*, 169–180. [CrossRef] [PubMed]
56. Mosher, C.E.; Lipkus, I.; Sloane, R.; Snyder, D.C.; Lobach, D.F.; Demark-Wahnefried, W. Long-Term Outcomes of the FRESH START Trial: Exploring the Role of Self-Efficacy in Cancer Survivors' Maintenance of Dietary Practices and Physical Activity. *Psychooncology* **2013**, *22*, 876–885. [CrossRef] [PubMed]
57. Huang, F.F.; Yang, Q.; Wang, A.N.; Zhang, J.P. Psychometric Properties and Performance of Existing Self-Efficacy Instruments in Cancer Populations: A Systematic Review. *Health Qual. Life Outcomes* **2018**, *16*, 1–12. [CrossRef] [PubMed]
58. Náfrádi, L.; Nakamoto, K.; Schulz, P.J. Is Patient Empowerment the Key to Promote Adherence? A Systematic Review of the Relationship between Self-Efficacy, Health Locus of Control and Medication Adherence. *PLoS ONE* **2017**, *12*, 1–23. [CrossRef] [PubMed]
59. Lettieri, E.; Fumagalli, L.P.; Radaelli, G.; Bertele, P.; Vogt, J.; Hammerschmidt, R.; Lara, J.L.; Carriazo, A.; Masella, C. Empowering Patients through EHealth: A Case Report of a Pan-European Poject Health Policy, Reform, Governance and Law. *BMC Health Serv. Res.* **2015**, *15*, 309. [CrossRef] [PubMed]
60. Pomerleau, J.; Lock, K.; Knai, C.; McKee, M. Interventions Designed to Increase Adult Fruit and Vegetable Intake Can Be Effective: A Systematic Review of the Literature. *J. Nutr.* **2005**, *135*, 2486–2495. [CrossRef] [PubMed]
61. Norman, G.R.; Sloan, J.A.; Wyrwich, K.W. Interpretation of Changes in Health-Related Quality of Life: The Remarkable Universality of Half a Standard Deviation. *Med. Care* **2003**, *41*, 582–592. [CrossRef]
62. Baguley, B.J.; Bolam, K.A.; Wright, O.R.L.; Skinner, T.L. The Effect of Nutrition Therapy and Exercise on Cancer-Related Fatigue and Quality of Life in Men with Prostate Cancer: A Systematic Review. *Nutrients* **2017**, *9*, 1003. [CrossRef]
63. Uster, A.; Ruehlin, M.; Mey, S.; Gisi, D.; Knols, R.; Imoberdorf, R.; Pless, M.; Ballmer, P.E. Effects of Nutrition and Physical Exercise Intervention in Palliative Cancer Patients: A Randomized Controlled Trial. *Clin. Nutr.* **2018**, *37*, 1202–1209. [CrossRef]
64. Grassi, L.; Caruso, R.; Mitchell, A.J.; Sabato, S.; Nanni, M.G. Screening for Emotional Disorders in Patients with Cancer Using the Brief Symptom Inventory (BSI) and the BSI-18 versus a Standardized Psychiatric Interview (the World Health Organization Composite International Diagnostic Interview). *Cancer* **2018**, *124*, 2415–2426. [CrossRef]
65. Kuhnt, S.; Brähler, E.; Faller, H.; Härter, M.; Keller, M.; Schulz, H.; Wegscheider, K.; Weis, J.; Boehncke, A.; Hund, B.; et al. Twelve-Month and Lifetime Prevalence of Mental Disorders in Cancer Patients. *Psychother. Psychosom.* **2016**, *85*, 289–296. [CrossRef] [PubMed]
66. Mitchell, A.J.; Chan, M.; Bhatti, H.; Halton, M.; Grassi, L.; Johansen, C.; Meader, N. Prevalence of Depression, Anxiety, and Adjustment Disorder in Oncological, Haematological, and Palliative-Care Settings: A Meta-Analysis of 94 Interview-Based Studies. *Lancet. Oncol.* **2011**, *12*, 160–174. [CrossRef]

67. Mehnert, A.; Hartung, T.J.; Friedrich, M.; Vehling, S.; Brähler, E.; Härter, M.; Keller, M.; Schulz, H.; Wegscheider, K.; Weis, J.; et al. One in Two Cancer Patients Is Significantly Distressed: Prevalence and Indicators of Distress. *Psychooncology* **2018**, *27*, 75–82. [CrossRef] [PubMed]
68. Bortolato, B.; Hyphantis, T.N.; Valpione, S.; Perini, G.; Maes, M.; Morris, G.; Kubera, M.; Köhler, C.A.; Fernandes, B.S.; Stubbs, B.; et al. Depression in Cancer: The Many Biobehavioral Pathways Driving Tumor Progression. *Cancer Treat. Rev.* **2017**, *52*, 58–70. [CrossRef]
69. Arnold, E.M. The Cessation of Cancer Treatment as a Crisis. *Soc Work Health Care.* **2008**, *1389*, 37–41. [CrossRef]
70. Jacka, F.N.; O'Neil, A.; Opie, R.; Itsiopoulos, C.; Cotton, S.; Mohebbi, M.; Castle, D.; Dash, S.; Mihalopoulos, C.; Chatterton, M.L.; et al. A Randomised Controlled Trial of Dietary Improvement for Adults with Major Depression (the "SMILES" Trial). *BMC Med.* **2017**, *15*, 1–13. [CrossRef]
71. Giese-Davis, J.; Brandelli, Y.; Kronenwetter, C.; Golant, M.; Cordova, M.; Twirbutt, S.; Chang, V.; Kraemer, H.C.; Spiegel, D. Illustrating the Multi-Faceted Dimensions of Group Therapy and Support for Cancer Patients. *Healthcare* **2016**, *4*, 48. [CrossRef]
72. American Cancer Society. *Cancer Treatment & Survivorship: Facts & Figures 2019–2021*; American Cancer Society: Atlanta, GA, USA, 2019.

Publisher's Note: MDPI stays neutral with regard to jurisdictional claims in published maps and institutional affiliations.

nutrients

MDPI

Article

Impact of Clinical Markers of Nutritional Status and Feeding Jejunostomy Use on Outcomes in Esophageal Cancer Patients Undergoing Neoadjuvant Chemoradiotherapy

Rishi Jain [1,*], Talha Shaikh [2], Jia-Llon Yee [1], Cherry Au [1], Crystal S. Denlinger [1], Elizabeth Handorf [3], Joshua E. Meyer [2] and Efrat Dotan [1]

1 Fox Chase Cancer Center, Department of Hematology/Oncology, Philadelphia, PA 19111-2497, USA; jiallony@gmail.com (J.-L.Y.); Cherry.au.001@gmail.com (C.A.); crystal.denlinger@fccc.edu (C.S.D.); efrat.dotan@fccc.edu (E.D.)

2 Fox Chase Cancer Center, Department of Radiation Oncology, Philadelphia, PA 19111-2497, USA; talha.s019@gmail.com (T.S.); joshua.meyer@fccc.edu (J.E.M.)

3 Biostatistics, Fox Chase Cancer Center, Philadelphia, PA 19111-2497, USA; elizabeth.handorf@fccc.edu

* Correspondence: rishi.jain@fccc.edu; Tel.: 215-728-2820

Received: 28 August 2020; Accepted: 15 October 2020; Published: 17 October 2020

Abstract: Background: Patients with esophageal cancer (EC) have high rates of malnutrition due to tumor location and treatment-related toxicity. Various strategies are used to improve nutritional status in patients with EC including oral and enteral support. Methods: We conducted a retrospective analysis to determine the impact of malnutrition and prophylactic feeding jejunostomy tube (FJT) placement on toxicity and outcomes in patients with localized EC who were treated with neoadjuvant chemoradiation therapy (nCRT) followed by esophagectomy. Results: We identified 125 patients who were treated with nCRT between 2002 and 2014. Weight loss and hypoalbuminemia occurred frequently during nCRT and were associated with multiple adverse toxicity outcomes including hematologic toxicity, nonhematologic toxicity, grade ≥3 toxicity, and hospitalizations. After adjusting for relevant covariates including the specific nCRT chemotherapy regimen received and the onset of toxicity, there were no significant associations between hypoalbuminemia, weight loss, or FJT placement and relapse-free survival (RFS) or overall survival (OS). FJT placement was associated with less weight loss during nCRT ($p = 0.003$) but was not associated with reduced toxicity or improved survival. Conclusions: Weight and albumin loss during nCRT for EC are important factors relating to treatment toxicity but not RFS or OS. While pretreatment FJT placement may reduce weight loss, it may not impact treatment tolerance or survival.

Keywords: nutrition; esophageal cancer; neoadjuvant chemoradiation; enteral nutrition

1. Introduction

In 2020, approximately 18,440 new cases of esophageal cancer (EC) will be diagnosed in the United States and 16,170 will die from this disease [1]. Unfortunately, despite advances in EC therapy, under 20% of patients survive for five years after diagnosis [1]. Up to 80% of EC patients present with pretreatment malnutrition, largely due to mechanical obstruction [2]. Furthermore, EC is associated with a hyperinflammatory, cachectic state marked by weight loss and sarcopenia [3]. The combination of nutrient imbalance and cancer cachexia leads to loss of muscle mass and resultant loss of physical functioning, reduced quality of life, and increased risk of treatment-related toxicity [3].

Curative intent therapy in EC involves neoadjuvant chemoradiotherapy (nCRT) followed by esophagectomy [4]. However, nCRT carries the risk of significant gastrointestinal (GI) toxicities, including any-grade nausea (53%), vomiting (25%), diarrhea (18%), constipation (27%), and esophagitis (19%), among others [5]. Thus, malnutrition can be further aggravated by the toxicity of nCRT leading to treatment delays or interruptions and worse outcomes [6]. While clinicians use changes in weight or serum albumin levels, consensus is lacking on the most reliable indicator of malnutrition. Commonly used strategies to treat malnutrition in EC may include oral nutritional supplements, enteral nutrition with a percutaneous endoscopic gastrostomy (PEG) or feeding jejunostomy tube (FJT), or even total parenteral nutrition (TPN) in severe cases [7]. Many institutions use preventive placement of enteral feeding support before the initiation of nCRT for EC [8].

While data support the use of postoperative enteral support after esophagectomy [9], limited data are available regarding the optimal method of preoperative nutritional support. While one study of oral nutritional intervention showed that a dietitian-delivered, preoperative nutritional intervention was able to reduce weight loss and decrease surgical morbidity [10], other studies using oral-based interventions have failed to show consistent benefits in outcomes such as nCRT tolerance [11]. The utility of preoperative enteral nutrition in EC has recently been questioned [12]. To date, there is no consensus regarding the optimal management of malnutrition in EC during nCRT. Further, there is little data available regarding the impact of malnutrition occurring during nCRT on outcomes. We conducted a single-center retrospective analysis evaluating the associations between markers of nutritional status and FJT placement on the tolerance and outcomes of patients with localized EC treated with nCRT.

2. Materials and Methods

2.1. Patient Selection and Data Collection

The study was conducted in accordance with the Declaration of Helsinki, and the protocol was approved by the Institutional Review Board (IRB) of Fox Chase Cancer Center (IRB# 15-9020) on 28 August 2014. Following IRB approval, we retrospectively identified patients with localized EC who underwent nCRT followed by esophagectomy between 2002 and 2014. The tumor registry was utilized to identify eligible patients. The following baseline data at the time of EC diagnosis were collected: age, gender, Eastern Cooperative Oncology Group (ECOG) performance status, comorbidities and Charlson comorbidity index (CCI), smoking status, tumor location, tumor histology, and clinical stage at diagnosis. Former smokers were defined as those who quit >1 year before diagnosis. Data regarding chemotherapy regimen used during radiation, total radiation dose, and pathologic stage were collected.

A detailed pre-nCRT and post-nCRT assessment of nutritional status was conducted for the following factors: placement of pretreatment FJT, weight change during nCRT (lbs), and albumin change during nCRT (g/dL). Pre-nCRT weight and albumin levels were within two weeks of initiation of CRT, while post-nCRT weight and albumin were obtained within one week of nCRT completion. For each patient, weight change was divided into the following categories: (1) gain of body weight, (2) loss of <5% of body weight, (3) loss of 5–10% of body weight, or (4) loss of > 10% of body weight. Albumin change was categorized into either (1) decrease of < 0.5 g/dL or (2) a decrease of ≥0.5 g/dL.

A comprehensive review of clinic notes and laboratory values before, during, and immediately after the completion of nCRT was conducted to identify the onset of toxicity. All toxicity endpoints are defined in Table 1. The onset of hematologic or renal toxicity was scored based on grade and severity per the Common Terminology Criteria for Adverse Events (CTCAE version 4.0). The onset of any-grade nonhematologic toxicity during nCRT was also recorded. Nonhematologic toxicities recorded included the onset of the following: mucositis, nausea, vomiting, diarrhea, constipation, anorexia, dehydration, fatigue, esophagitis, dysphagia, or infections. Due to limitations in clinical documentation, the presence of nonhematologic toxicity endpoints was not graded but rather recorded as binary (yes or no). If a hematologic or nonhematologic toxicity occurred multiple times in the same patient, only one event was included for analysis. We also collected data on any nCRT-related hospitalizations, chemotherapy dose interruptions, or dose reductions. All data collection was completed by two reviewers to ensure accuracy.

Table 1. Definitions of toxicity outcomes.

Hospitalizations	Any Unexpected Hospitalization during nCRT
Nonhematologic toxicity [1]	Any-grade mucositis, nausea, vomiting, diarrhea, constipation, anorexia, dehydration, fatigue, esophagitis, dysphagia, neurotoxicity
Hematologic toxicity [2]	Neutropenia, thrombocytopenia, anemia
Grade ≥3 toxicity [3]	Onset of any grade ≥3 toxicity during nCRT Hgb < 8 g/dL, ANC < 1000/mm3, Plts < 50,000/mm3, Cr < 3× baseline or >3.0 mg/dL
Dose reduction or interruption [4]	Unplanned chemotherapy dose reduction or interruption

nCRT, neoadjuvant chemoradiation therapy; Hgb, hemoglobin; ANC, absolute neutrophil count; Plts, platelets; Cr, creatinine. [1] Includes any-grade nonhematologic toxicity. [2] Weighted scores given to higher grade hematologic toxicity. [3] Grade ≥3 toxicity including hematologic or renal insufficiency. [4] Any dose reduction or interruption regardless of percent of reduction or timing of interruption.

2.2. Statistical Analysis

Univariate analyses (UVA) were used to determine associations between covariates and nutritional outcomes, using Fisher's exact test to determine significance of relationships. Multivariable regression analyses (MVA) were used to determine associations between nutritional markers and toxicity, adjusting for the following covariates: age, gender, smoking status, pathologic stage, nCRT regimen, ECOG performance status, and comorbidities (CCI). Linear regression analyses were used to model hematologic and nonhematologic toxicity scores, and logistic regression analyses were used to model the presence or absence of dose reductions/interruptions, hospitalizations, and grade ≥3 toxicities. Multivariable Cox proportional hazards regression models were also used to determine associations between nutritional markers and relapse-free survival (RFS) and overall survival (OS) adjusting for the following covariates: onset of any grade ≥3 toxicity during nCRT, the chemotherapy regimen chosen during nCRT, and also the presence of a pathologic complete response after nCRT. All analyses were performed using Stata software (version 12) StataCorp LLC (College Station, TX, USA), and p-values < 0.05 were considered statistically significant.

3. Results

3.1. Baseline Characteristics and Markers of Nutritional Status

A total of 125 patients who underwent nCRT before esophagectomy for EC were identified. Baseline demographics and disease and treatment characteristics are shown in Table 2. The patients had a median age of 63 and were primarily male (84%). The majority of patients had lower esophageal tumors (88%) and adenocarcinoma (86%). The majority of patients had no comorbidities with a CCI of 0 (70%) and had an ECOG performance score of 0 (56%). Clinical stage III EC was most common (50%), while the most common pathologic stage following esophagectomy was stage II (38%). Pathologic complete response was achieved in 32 patients (26%). The most common chemotherapy regimen used was cisplatin/5-FU in 65 patients (52%), while 45 patients (36%) received carboplatin and paclitaxel. The majority of patients (74%) received ≥5000cGy of total radiation.

Table 2. Patient tumor and treatment characteristics.

Characteristic	Total N = 125 (%)
Median Age	63, range 35–80
Gender	
Male	105 (84)
Female	20 (16)
Pathologic subtype	
Adenocarcinoma	108 (86)
Squamous	17 (14)
Location	
Lower tumor	110 (88)
Other	15 (12)
Smoking status	
Past	71 (57)
Current	21 (17)
Former	33 (26)
Charlson comorbidity index	
CCI = 0	87 (70)
CCI = 1	25 (20)
CCI >1	13 (11)
Performance status	
0	70 (56)
≥1 or greater	55 (44)
Clinical stage	
I	1 (0)
II	48 (38)
III	63 (50)
IV [1]	13 (10)
Pathologic stage	
0 [2]	32 (26)
I	15 (12)
II	47 (38)
III	24 (19)
IV [3]	7 (6)
Chemotherapy regimen	
Carboplatin/Paclitaxel	45 (36)
Cisplatin/5-FU	65 (52)
Other [4]	15 (12)
Radiation dose	
<5000 cGy	33 (26)
≥5000 cGy	92 (74)

CCI, Charlson comorbidity index;. [1] Patients with clinical stage IV disease most often had locally advanced lymphadenopathy that was able to be covered by radiation field. [2] Pathologic stage was determined after esophagectomy, pathologic stage 0 indicates a pathologic complete response. [3] Five were clinical stage IV, two were upstaged at time of surgery. [4] Nine treated on clinical trial with combination of Vandetanib, Paclitaxel, Carboplatin, 5-FU.

3.2. *Markers of Malnutrition and Toxicity during nCRT*

Nutritional characteristics before and after nCRT are shown in Table 3. Mean weight loss during nCRT completion was 10 lbs, with the most common category of weight loss being 5–10% occurring in 36% of patients. Mean albumin decrease during nCRT was 0.4 g/dL. There was an even distribution of patients with or without a pretreatment FJT (62 and 63 patients, respectively). Patients with an FJT placed prior to nCRT had a median year of diagnosis of 2011, while those without FJT placement had a median year of diagnosis of 2007. After nCRT completion, mean weight loss was significantly reduced in patients with an FJT vs. those without an FJT (8 lbs vs. 13 lbs, p = 0.003). However, no significant difference in mean albumin loss was noted in those with an FJT vs. those without FJT (0.38 g/dL vs. 0.52 g/dL, p = 0.15). The prevalence of nCRT-related toxicity is also shown in

Table 3. Chemotherapy dose reductions were required in 10% of patients while dose interruptions were required in 22%. In total, 26% of patients had dose interruptions or reductions and were unable to receive all planned chemotherapy. Eighteen percent of patients had unplanned hospitalizations during nCRT for management of nCRT-related toxicity.

Table 3. Changes in nutritional markers and onset of toxicity during neoadjuvant chemoradiation therapy.

Characteristic	Total $N = 125$ (%)
Mean weight pre-nCRT (lbs)	188
Mean weight post-nCRT	178
Mean weight change	10
Weight change category	
Weight gain	20 (16)
Weight loss <5%	40 (32)
Weight loss ≥5–10%	45 (36)
Weight loss ≥10%	20 (16)
Mean albumin [1] (pre-nCRT) (g/dL)	3.8
Mean albumin (post-nCRT) (g/dL)	3.4
Mean albumin (g/dL) loss	0.4
Albumin change	
<0.5 g/dL	59 (47)
≥0.5 g/dL	61 (49)
Pre-CRT feeding jejunostomy tube placement	
Yes	62 (50)
No	63 (50)
Mean weight loss with FJT (lbs)	8
Mean weight loss without FJT (lbs)	13
Mean albumin loss with FJT (g/dL)	0.38
Mean albumin loss without FJT (g/dL)	0.52
Chemotherapy dose reductions	12 (10)
Chemotherapy dose interruptions	27 (22)
Dose reductions/interruptions	33 (26)
Hospitalizations	23 (18)

nCRT, neoadjuvant chemoradiation therapy; FJT, feeding jejunostomy tube; [1] Albumin levels were unavailable in 4 patients.

Associations between baseline and treatment characteristics and markers of malnutrition are shown in Table S1. Patients treated with nCRT with cisplatin/5-FU experienced higher percent weight loss compared to those treated with carboplatin/paclitaxel ($p < 0.001$). Patients who received carboplatin/paclitaxel were more likely to have an FJT (84% in carboplatin/paclitaxel versus 31% in cisplatin/5-FU, $p < 0.001$). Patients with an ECOG performance score of 1 were more likely to have an FJT than those with a score of 0 ($p = 0.025$). Otherwise, there were no significant associations between markers of malnutrition or FJT placement and age, gender, smoking status, pathologic stage, or comorbidity score.

3.3. *Association between Albumin, Weight Loss, and Toxicity during CRT*

The results of the UVA and MVA are shown in Table 4. Various markers of malnutrition were associated with toxicity. An albumin loss of ≥0.5 during nCRT was significantly associated with the occurrence of nonhematologic toxicity, hematologic toxicity, and any grade ≥3 toxicity in UVA. After adjusting for covariates in the MVA, an albumin loss of ≥0.5 during nCRT was significantly associated with hospitalizations, nonhematologic toxicity, hematologic toxicity, and any grade ≥3 toxicity. When viewing specific nonhematologic toxicities and albumin change, higher rates of mucositis (61% vs. 36%), nausea (79% vs. 63%), vomiting (47% vs. 37%), diarrhea (30% vs. 24%), and anorexia (52% vs. 37%) were noted in those with an albumin loss of ≥ 0.5 versus albumin loss of <0.5. A weight loss of 5–10% and >10% during nCRT was significantly associated with nonhematologic toxicity in UVA and

MVA, while a weight loss of >10% was significantly associated with higher grade ≥3 toxicity in the MVA. There were no associations between pretreatment FJT status and toxicity rates in UVA or MVA.

Table 4. Associations between nutritional markers and toxicity.

	Albumin Change (<0.5 vs. ≥0.5)		Weight Loss (Gain or <5%, 5–10%, >10%)		J-Tube (Yes vs. No)	
Toxicity	UVA (*p*-value)	MVA ($p > z$ or $p > t$)	UVA (*p*-value)	MVA ($p > z$ or $p > t$)	UVA (*p*-value)	MVA ($p > z$ or $p > t$)
Dose reductions/interruptions	NS	0.043	NS	NS	NS	NS
Hospitalizations	NS	NS	NS	NS	NS	NS
Nonhematologic toxicity	0.011	0.004	0.019	<5% = NS 5–10% = 0.035 >10% = 0.002	NS	NS
Hematologic toxicity	0.004	0.002	NS	NS	NS	NS
Grade ≥3 toxicity	0.006	0.004	NS	<5% = NS 5–10% = NS >10% = 0.032	NS	NS

UVA, univariate analysis; MVA, multivariable regression analyses; NS, nonsignificant.

3.4. RFS and OS by Albumin, Weight Loss, and FJT Status

The results of the MVA of RFS and OS by nutritional markers or FJT placement status are shown in Table 5. After adjusting for relevant covariates including the onset of grade ≥ 3 toxicity, the specific chemotherapy regimen used during nCRT, and the presence of a complete response to neoadjuvant therapy, there were no significant associations between FJT status, albumin change, or weight loss and RFS or OS.

Table 5. Associations between nutritional markers and relapse-free and overall survival.

	Albumin Change (<0.5 vs. ≥0.5)	Weight Loss (<5% vs. ≥5%)	J-Tube (Yes vs. No)
	MVA HR [95% CI], *p*-Value	**MVA HR [95% CI], *p*-Value**	**MVA HR [95% CI], *p*-Value**
Relapse-free survival	1.13 [0.64–2.01], $p = 0.67$	1.29 [0.71–2.32], $p = 0.40$	1.36 [0.69–2.69], $p = 0.37$
Overall survival	0.98 [0.57–1.70], $p = 0.95$	1.27 [0.73–2.24], $p = 0.40$	0.86 [0.47–1.59], $p = 0.63$

MVA, multivariable regression analyses; HR, hazard ratio; CI, confidence interval.

4. Discussion

Pretreatment malnutrition is a common occurrence in patients with localized EC and may be exacerbated by the toxicity of nCRT used as part of standard trimodality therapy (chemoradiation therapy followed by esophagectomy) [4,6]. As malnutrition and associated muscle loss have been strongly associated with multiple adverse treatment outcomes including increased chemotherapy toxicity [13] and reduced survival [14] in other malignancies, deterioration of nutritional status in patients undergoing nCRT for EC may be detrimental and further compromise survival in this deadly disease. While postesophagectomy nutritional interventions such as FJT placement are beneficial [15], there is little data on the impact of changes in nutritional markers during nCRT on outcomes for EC.

We conducted a single-center retrospective analysis of patients with EC who underwent nCRT prior to esophagectomy. The majority of EC patients either developed or had exacerbation of weight loss and hypoalbuminemia during nCRT. Our data are in line with prior studies showing high rates of malnutrition related to nCRT [16]. Higher degrees of hypoalbuminemia or weight loss during nCRT were associated with significantly higher rates of multiple important toxicity measures, including chemotherapy dose modifications and nonhematologic, hematologic, and grade ≥3 toxicity. Associations between impaired treatment tolerance and malnutrition or sarcopenia have previously been identified with a variety of malignancies and cancer therapies [17,18]. While not thoroughly studied, potential hypotheses linking these conditions include acquired pharmacokinetic differences in drug metabolism in patients with low lean muscle mass, variations of fat-free mass and volume of distribution, among others [19]. Patients with profound weight and muscle loss during nCRT may be unable to receive as much chemotherapy due to missed or reduced chemotherapy doses in the setting of toxicity such as renal dysfunction or cytopenias. Notably, nonhematologic toxicities such as mucositis, nausea, and vomiting were more common in patients with more pronounced hypoalbuminemia. These challenging symptoms are mostly attributed to chemotherapy and may also have necessitated dose modifications resulting in inadequate therapy. This warrants further investigation.

Weight loss was also associated with the type of chemotherapy regimen utilized, with higher rates of weight loss in patients treated with cisplatin/5-FU compared to the more contemporary regimen of carboplatin/paclitaxel. This is likely related to higher GI toxicity, particularly nausea and mucositis, which have been associated with cisplatin/5-FU in clinical trials [20]. Malnutrition in patients undergoing nCRT is likely further compounded by toxicity associated with radiotherapy, especially esophagitis and dysphagia. Some data suggest that neoadjuvant chemotherapy prior to nCRT may reduce malnutrition due to less dysphagia, resultant weight gain, and improved quality of life [21]. More data regarding this approach and its impact on nutritional status and other outcomes are necessary.

Beyond nutritional consultation and use of oral nutritional supplements, enteral support with prophylactic FJT placement prior to nCRT is increasingly common at many centers. In our cohort, FJT placement was most common in patients treated with carboplatin and paclitaxel, reflecting changes in practice patterns. FJT placement was also more common in patients with a higher ECOG performance score, suggesting that reduced baseline functioning may have contributed to the decision for FJT placement. Although performance status is not typically a criterion for FJT use, physicians may choose more aggressive nutritional support mechanisms in those who are weaker or deconditioned. Despite a small reduction in weight loss (5 lbs), there was no reduction in toxicity rates in patients who had an FJT placed prior to nCRT. Other studies have also questioned the utility for preoperative enteral feeding. Studies have shown high rates of tube-related complications without a substantial improvement in nutritional parameters [12,22]. With complication rates up to 40% related to FJT use, including some even requiring operative interventions [23], a detailed risk/benefit discussion must take place with the patient prior to FJT placement. It is likely that certain select patients with critical malnutrition and severe obstruction may benefit most from PEG or FJT placement before nCRT. Given the lack of high-level evidence of enteral support during nCRT, consensus guidelines from major organizations do not mandate enteral support but recommend consideration of FJT placement in those with significant impairment of oral intake [24,25]. Additional prospective research regarding the utility and optimal timing of FJT placement is warranted.

There is little available data regarding changes in nutritional markers during nCRT and survival outcomes. To further understand these relationships in our cohort, we conducted an MVA of nutritional status and RFS/OS after adjusting for relevant covariates including the onset of grade ≥3 toxicity, the specific chemotherapy regimen used during nCRT, and the presence of a complete pathologic response to nCRT, a known prognostic marker. Despite the strong associations between weight loss and albumin loss and key toxicity endpoints in our UVA and MVA, there were no significant associations between weight loss and hypoalbuminemia and RFS/OS. Furthermore, there was no impact of FJT status on survival outcomes. While short-term changes in nutritional status during the 5–6 weeks of nCRT may not impact survival outcomes, additional data are needed to understand the effect of longer-term changes in nutritional markers on survival outcomes. Efforts should be made to identify better biomarkers of nutritional status which may be more reflective of the true burden of malnutrition and the resultant catabolic physiologic state (e.g., markers of inflammation) in this population.

There are limitations of our study. As a single-institution analysis, additional data from other centers would help us understand the impact of nutritional status on toxicity and outcomes in larger populations. The retrospective nature of our study makes it challenging to determine causality of malnutrition with adverse outcomes as other patient, tumor, or treatment-related factors may have varying degrees of contribution. Retrospective evaluation of nutritional status and treatment-related toxicity is challenging given limitations in clinical documentation. To overcome these challenges, each chart was reviewed independently by two reviewers to assess accuracy. Finally, the diagnosis of esophageal cancer in our cohort ranged over a period of 12 years, a time frame when the standard of care for nCRT for localized esophageal cancer evolved as did nutritional support methods and supportive care capabilities. Despite these limitations, given the high morbidity of nCRT and poor survival of patients with esophageal cancer, these preliminary data support ongoing efforts to improve nutritional status during nCRT in an effort to reduce toxicity and improve outcomes.

In the future, assessment of muscle status in conjunction with markers of malnutrition may be useful in selecting patients who would benefit the most from aggressive nutritional support. In those with evidence of severe muscle depletion and systemic inflammation, addition of one of the many anticachexia agents in development in conjunction with nutritional support may be especially advantageous [26]. In cases of severe malnutrition from esophageal tumor obstruction, esophageal stent placement is increasingly used in the palliative setting and can lead to an immediate improvement of obstructive symptoms. While this may prevent the need for enteral nutrition, risks including esophageal fistula formation have been reported when stents were placed prior to chemoradiotherapy [27]. While these strategies hold promise, further prospective evaluation is necessary before standard use in clinical settings.

In this study, we demonstrate strong associations between malnutrition and increased toxicity but not survival among patients with localized EC treated with nCRT. There was no improvement in outcomes with FJT use, which calls into question the utility of prophylactic FJT placement prior to nCRT. Our findings emphasize the need for future research to identify novel strategies to reduce malnutrition and improve the tolerance of nCRT. Future studies will require multidisciplinary strategies—potentially incorporating nutritional support, physical activity, and novel anti-inflammatory/anticachexia pharmacotherapies—with the goal of improved treatment tolerance, quality of life, and, ultimately, survival.

Supplementary Materials: The following are available online at http://www.mdpi.com/2072-6643/12/10/3177/s1, Table S1: Associations between patient characteristics and markers of nutritional status.

Author Contributions: Conceptualization, R.J., E.D. and J.E.M.; methodology, R.J. and E.H.; software, E.H.; validation, R.J. and E.D.; formal analysis, R.J. and E.H.; investigation, R.J. and E.D.; resources, E.D. and J.E.M.; data curation, R.J., T.S., J.-L.Y., and C.A.; writing—original draft preparation, R.J.; writing—review and editing, R.J., T.S., J.-L.Y., C.A., E.H., C.S.D., J.E.M. and E.D.; visualization, R.J. and E.H.; supervision, E.D.; project administration, R.J. and E.D.; funding acquisition, E.D. All authors have read and agreed to the published version of the manuscript.

Funding: Supported by Cancer Center Support Grant 3 P30 CA006927-47S4.

Conflicts of Interest: The authors declare no conflict of interest. The funders had no role in the design of the study; in the collection, analyses, or interpretation of data; in the writing of the manuscript, or in the decision to publish the results.

References

1. Esophageal Cancer—Cancer Stat Facts. Available online: https://seer.cancer.gov/statfacts/html/esoph.html (accessed on 16 October 2020).
2. Riccardi, A. Nutritional Management of Patients with Esophageal and Esophagogastric Junction Cancer. *Cancer Control.* **1999**, *6*, 64–72. [CrossRef] [PubMed]
3. Anandavadivelan, P.; Lagergren, P. Cachexia in patients with oesophageal cancer. *Nat. Rev. Clin. Oncol.* **2016**, *13*, 185–198. [CrossRef] [PubMed]
4. Sjoquist, K.M.; Burmeister, B.H.; Smithers, B.M.; Zalcberg, J.R.; Simes, R.J.; Barbour, A.; Gebski, V.; Australasian Gastro-Intestinal Trials Group. Survival after neoadjuvant chemotherapy or chemoradiotherapy for resectable oesophageal carcinoma: An updated meta-analysis. *Lancet Oncol.* **2011**, *12*, 681–692. [CrossRef]
5. van Hagen, P.; Hulshof, M.C.C.M.; van Lanschot, J.J.B.; Steyerberg, E.W.; Henegouwen, M.V.B.; Wijnhoven, B.P.L.; Richel, D.J.; Nieuwenhuijzen, G.A.P.; Hospers, G.A.P.; Bonenkamp, J.J.; et al. Preoperative chemoradiotherapy for esophageal or junctional cancer. *N. Engl. J. Med.* **2012**, *366*, 2074–2084. [CrossRef]

6. Clavier, J.B.; Antoni, D.; Atlani, D.; Ben Abdelghani, M.; Schumacher, C.; Dufour, P.; Kurtz, J.E.; Noel, G. Baseline nutritional status is prognostic factor after definitive radiochemotherapy for esophageal cancer. *Dis. Esophagus* **2014**, *27*, 560–567. [CrossRef]
7. Birnstein, E.; Schattner, M. Nutritional Support in Esophagogastric Cancers. *Surg. Oncol. Clin. N. Am.* **2017**, *26*, 325–333. [CrossRef]
8. Margolis, M.; Alexander, P.; Trachiotis, G.D.; Gharagozloo, F.; Lipman, T. Percutaneous endoscopic gastrostomy before multimodality therapy in patients with esophageal cancer. *Ann. Thorac. Surg.* **2003**, *76*, 1694–1698. [CrossRef]
9. Couper, G. Jejunostomy after oesophagectomy: A review of evidence and current practice. *Proc. Nutr. Soc.* **2011**, *70*, 316–320. [CrossRef] [PubMed]
10. Ligthart-Melis, G.C.; Weijs, P.J.M.; te Boveldt, N.D.; Buskermolen, S.; Earthman, C.P.; Verheul, H.M.W.; de Lange-de Klerk, E.S.M.; van Weyenberg, S.J.B.; van der Peet, D.L. Dietician-delivered intensive nutritional support is associated with a decrease in severe postoperative complications after surgery in patients with esophageal cancer. *Dis. Esophagus* **2013**, *26*, 587–593. [CrossRef]
11. Sikora, S.S.; Ribeiro, U.; Kane, J.M.; Landreneau, R.J.; Lembersky, B.; Posner, M.C. Role of nutrition support during induction chemoradiation therapy in esophageal cancer. *J. Parenter. Enteral Nutr.* **1998**, *22*, 18–21. [CrossRef] [PubMed]
12. Jenkins, T.K.; Lopez, A.N.; Sarosi, G.A.; Ben-David, K.; Thomas, R.M. Preoperative enteral access is not necessary prior to multimodality treatment of esophageal cancer. *Surgery* **2018**, *163*, 770–776. [CrossRef] [PubMed]
13. Prado, C.M.M.; Maia, Y.L.M.; Ormsbee, M.; Sawyer, M.B.; Baracos, V.E. Assessment of nutritional status in cancer-the relationship between body composition and pharmacokinetics. *Anticancer Agents Med. Chem.* **2013**, *13*, 1197–1203. [CrossRef] [PubMed]
14. Blauwhoff-Buskermolen, S.; Versteeg, K.S.; de van der Schueren, M.A.; Den Braver, N.R.; Berkhof, J.; Langius, J.A.; Verheul, H.M. Loss of Muscle Mass During Chemotherapy Is Predictive for Poor Survival of Patients with Metastatic Colorectal Cancer. *J. Clin. Oncol.* **2016**, *34*, 1339–1344. [CrossRef]
15. Peng, J.; Cai, J.; Niu, Z.X.; Chen, L.Q. Early enteral nutrition compared with parenteral nutrition for esophageal cancer patients after esophagectomy: A meta-analysis. *Dis. Esophagus* **2016**, *29*, 333–341. [CrossRef] [PubMed]
16. Adenis, A.; Tresch, E.; Dewas, S.; Romano, O.; Messager, M.; Amela, E.; Clisant, S.; Kramar, A.; Mariette, C.; Mirabel, X. Clinical complete responders to definite chemoradiation or radiation therapy for oesophageal cancer: Predictors of outcome. *BMC Cancer* **2013**, *13*, 413. [CrossRef]
17. Klute, K.A.; Brouwer, J.; Jhawer, M.; Sachs, H.; Gangadin, A.; Ocean, A.; Popa, E.; Dai, T.; Wu, G.; Christos, P.; et al. Chemotherapy dose intensity predicted by baseline nutrition assessment in gastrointestinal malignancies: A multicentre analysis. *Eur. J. Cancer* **2016**, *63*, 189–200. [CrossRef] [PubMed]
18. Prado, C.M.M.; Baracos, V.E.; McCargar, L.J.; Reiman, T.; Mourtzakis, M.; Tonkin, K.; Mackey, J.R.; Koski, S.; Pituskin, E.; Sawyer, M.B. Sarcopenia as a Determinant of Chemotherapy Toxicity and Time to Tumor Progression in Metastatic Breast Cancer Patients Receiving Capecitabine Treatment. *Clin. Cancer Res.* **2009**, *15*, 2920–2926. [CrossRef]
19. Hopkins, J.J.; Sawyer, M.B. A review of body composition and pharmacokinetics in oncology. *Expert Rev. Clin. Pharmacol.* **2017**, *10*, 947–956. [CrossRef]
20. Honing, J.; Smit, J.K.; Muijs, C.T.; Burgerhof, J.G.M.; de Groot, J.W.; Paardekooper, G.; Muller, K.; Woutersen, D.; Legdeur, M.J.C.; Fiets, W.E.; et al. A comparison of carboplatin and paclitaxel with cisplatinum and 5-fluorouracil in definitive chemoradiation in esophageal cancer patients. *Ann. Oncol.* **2014**, *25*, 638–643. [CrossRef]
21. Cools-Lartigue, J.; Jones, D.; Spicer, J.; Zourikian, T.; Rousseau, M.; Eckert, E.; Alcindor, T.; Vanhuyse, M.; Asselah, J.; Ferri, L.E. Management of Dysphagia in Esophageal Adenocarcinoma Patients Undergoing Neoadjuvant Chemotherapy: Can Invasive Tube Feeding be Avoided? *Ann. Surg. Oncol.* **2015**, *22*, 1858–1865. [CrossRef]
22. Starr, B.; Davis, S.; Ayala-Peacock, D.; Blackstock, W.A.; Levine, E.A. Reassessment of the role of enteral tube feedings for patients with esophageal cancer. *Am. Surg.* **2014**, *80*, 752–758. [CrossRef] [PubMed]

23. Choi, A.H.; O'Leary, M.P.; Merchant, S.J.; Sun, V.; Chao, J.; Raz, D.J.; Kim, J.Y.; Kim, J. Complications of Feeding Jejunostomy Tubes in Patients with Gastroesophageal Cancer. *J. Gastrointest. Surg.* **2017**, *21*, 259–265. [CrossRef] [PubMed]
24. Arends, J.; Bachmann, P.; Baracos, V.; Barthelemy, N.; Bertz, H.; Bozzetti, F.; Fearon, K.; Hütterer, E.; Isenring, E.; Kaasa, S.; et al. ESPEN guidelines on nutrition in cancer patients. *Clin. Nutr.* **2017**, *36*, 11–48. [CrossRef]
25. Ajani, J.A.; D'Amico, T.A.; Almhanna, K.; Bentrem, D.J.; Besh, S.; Chao, J.; Das, P.; Denlinger, C.; Fanta, P.; Fuchs, C.S.; et al. Esophageal and esophagogastric junction cancers, version 1. 2015. *J. Natl. Compr. Cancer Netw.* **2015**, *13*, 194–227. [CrossRef] [PubMed]
26. Advani, S.M.; Advani, P.G.; VonVille, H.M.; Jafri, S.H. Pharmacological management of cachexia in adult cancer patients: A systematic review of clinical trials. *BMC Cancer* **2018**, *18*, 1174. [CrossRef]
27. Lu, Y.F.; Chung, C.S.; Liu, C.Y.; Shueng, P.W.; Wu, L.J.; Hsu, C.X.; Kuo, D.Y.; Hou, P.Y.; Chou, H.L.; Leong, K.I.; et al. Esophageal Metal Stents with Concurrent Chemoradiation Therapy for Locally Advanced Esophageal Cancer: Safe or Not? *Oncologist* **2018**, *23*, 1426–1435. [CrossRef] [PubMed]

nutrients

Review

Exercise and Nutrition Interventions in Patients with Head and Neck Cancer during Curative Treatment: A Systematic Review and Meta-Analysis

Asta Bye [1,2,*], Jon A. Sandmael [3], Guro B. Stene [4,5], Lene Thorsen [6,7], Trude R. Balstad [4,5], Tora S. Solheim [4,5], Are Hugo Pripp [1,8] and Line M. Oldervoll [9,10]

1 Faculty of Health Sciences, OsloMet—Oslo Metropolitan University, 0130 Oslo, Norway; apripp@oslomet.no
2 Regional Advisory Unit for Palliative Care, Department of Oncology, Oslo University Hospital, 0424 Oslo, Norway
3 Unicare Røros, Øverhagaen 15, 7374 Røros, Norway; jon.arne.sandmael@unicare.no
4 Department of Cancer Research and Molecular Medicine, Faculty of Medicine and Health Sciences, Norwegian University of Science and Technology (NTNU), 7491 Trondheim, Norway; guro.b.stene@ntnu.no (G.B.S.); trude.r.balstad@ntnu.no (T.R.B.); tora.s.solheim@ntnu.no (T.S.S.)
5 Cancer Clinic, St. Olavs Hospital, Trondheim University Hospital, 7006 Trondheim, Norway
6 National Advisory Unit on Late Effects after Cancer Treatment, Department of Oncology, Oslo University Hospital, 0424 Oslo, Norway; LKA@ous-hf.no
7 Department for Clinical Service, Division of Cancer Medicine, Oslo University Hospital, 0424 Oslo, Norway
8 Oslo Centre of Biostatistics and Epidemiology, Research Support Services, Oslo University Hospital, 0424 Oslo, Norway
9 Center for Crisis Psychology, Faculty of Psychology, University of Bergen, 5020 Bergen, Norway; line.oldervoll@ntnu.no
10 Department of Public Health and Nursing, Faculty of Medicine and Health Sciences, The Norwegian University of Science and Technology (NTNU), 7491 Trondheim, Norway
* Correspondence: abye@oslomet.no; Tel.: +47-97568595

Received: 11 September 2020; Accepted: 20 October 2020; Published: 22 October 2020

Abstract: The aim of this meta-analysis was to examine the effects of nutritional and physical exercise interventions and interventions combining these interventions during radiotherapy treatment for patients with head and neck cancer on body composition, objectively measured physical function and nutritional status. Systematic electronic searches were conducted in MEDLINE (PubMed interface), EMBASE (Ovid interface), CINAHL (EBSCO interface) and Cochrane Library (Wiley interface). We identified 13 randomized controlled trials (RCTs) that included 858 patients. For body composition, using only nutrition as intervention, a significant difference between treatment and control group were observed (SMD 0.42 (95CI 0.23–0.62), $p < 0.001$). Only pilot RCTs investigated combination treatment and no significant difference between the treatment and control groups were found (SMD 0.21 (95CI −0.16–0.58), $p = 0.259$). For physical function, a significant difference between treatment and control group with a better outcome for the treatment group were observed (SMD 0.78 (95CI 0.51–1.04), $p < 0.001$). No effects on nutritional status were found. This meta-analysis found significantly positive effects of nutrition and physical exercise interventions alone in favor of the treatment groups. No effects in studies with combined interventions were observed. Future full-scaled RCTs combining nutrition and physical exercise is warranted.

Keywords: head and neck neoplasms; nutrition; physical exercise; radiotherapy; meta-analysis; nutritional status; physical function; body composition

1. Introduction

Head and neck cancers (HNCs) comprises malignancies of the oral cavity, throat, larynx, salivary glands as well as nasal and paranasal sinuses. Surgery and radiotherapy (RT), sometimes combined with chemotherapy (CT) are the main treatment approaches [1]. Aggressive treatment regimens are effective to achieve tumor control and cure patients, but they also cause severe side-effects such as mouth dryness, mucositis and difficulties in swallowing [2]. Eating challenges due to tumor growth is one of the presenting symptoms of HNCs for many patients. Not surprisingly, when the challenges of the tumor is amplified by side effects of treatment that compromise dietary intake, many patients experience unintentional weight loss accompanied with muscle wasting [3]. Muscle wasting may influence muscle function and lead to loss of strength, increase fatigue and decrease quality of life [4].

To counteract the negative effects of weight loss and diminishing muscle mass for patients with HNCs during RT, it is recommended to ensure nutritional intake primarily through nutritional counseling and/or use of oral nutritional supplements (ONS) [5,6]. These recommendations are based on reviews indicating that dietary counselling can improve nutritional status and quality of life during RT [7,8]. However, the evidence supporting these strategies are inconclusive partly because previous reviews were not limited to randomized controlled trials (RCTs), patients with HNCs or interventions starting simultaneously with anticancer therapy. Physical exercise is another strategy that has the potential to decrease muscle catabolism and increase anabolism [5]. For patients with HNCs, exercise interventions have been tested in several pilot studies and are shown to be feasible, safe and to have potential impact on body composition, physical function, quality of life and fatigue management [9,10].

For patients treated for HNCs, weight loss, loss of strength, fatigue, and decreased quality of life are parts of a multidimensional problem related to both inadequate food intake and inactivity [7,11,12]. It is therefore a need to examine the impact of interventions combining nutrition and physical exercise as well as the feasibility of such interventions in this exposed population. Most previous studies have focused on either nutrition or physical exercise. However, physical exercise may be of importance for full effect of nutritional interventions and vice versa, sufficient nutrition is essential for optimal effect of physical exercise [5]. It could thus be hypothesized that a treatment approach including both nutrition and exercise is more effective improving patient outcomes than each intervention given alone. The aim of this systematic review and meta-analysis is therefore to examine current evidence for nutritional interventions alone, physical exercise interventions alone and interventions combining nutrition and physical exercise during radiotherapy treatment for patients with head and neck cancer. The main research questions are: (1) What are the effects on nutritional status, body composition and objectively measured physical function? (2) What is the content of the interventions? (3) What is adherence to and completion rate of the different interventions?

2. Material and Methods

The present review was conducted in accordance with the Preferred Reporting Items for Systematic Reviews and Meta Analyses (PRISMA) statement guidelines [13] and the review protocol is registered in the International Prospective Register of Systematic Reviews (PROSPERO) (Reg. nr.; CRD42018081487).

2.1. Data Sources, Search Strategy and Selection Criteria

2.1.1. Data Sources

Electronic searches were conducted the 31 October 2018 in MEDLINE (PubMed interface), EMBASE (Ovid interface), CINAHL (EBSCO interface) and Cochrane Library (Wiley interface). Additionally, the reference list of included studies and relevant systematic reviews were screened. Updated search in MEDLINE for the period between the 1 November 2018 and 3 June 2019 was later conducted to identify any additional relevant publication.

2.1.2. Search Strategy

The searches consisted of combinations of controlled terminology and free-text terms expressing the concepts (1) head and neck cancer and (2) exercise and (3) nutrition, adapted to each specific database. In Appendix A, the full search strategy of each database is described.

2.1.3. Selection Criteria

Inclusion criteria are shown in Table 1. Full scale RCTs and pilot RCTs evaluating the feasibility and/or effect of nutritional interventions and/or physical exercise published in peer-reviewed journals were considered for inclusion.

Table 1. PICOS (patients/population, intervention, comparator, outcomes, study design) criteria for inclusion and exclusion of studies.

	Inclusion Criteria	Exclusion Criteria
Population	Adults diagnosed with HNC, receiving RT with curative intent (± concomitant CT)	Patients <18 years of age, cancer with another origin, surgery as only treatment
Intervention	(1) Physical exercise or (2) nutrition or (3) a combination of exercise and nutrition. Initiated at start of RT and conducted during RT. Physical exercise is defined as sessions of muscle strength and/or aerobic exercise. Nutrition is defined as, dietary counselling, oral nutritional supplements or enteral nutrition by nasogastric tube or PEG	(1) Interventions initiated before start or after completion of RT (2) Nutritional interventions consisting only of vitamins or minerals (3) Comparisons of enteral and parenteral solutions (4) Swallowing exercise interventions alone
Comparator	Standard care or placebo	
Outcome	Nutritional status (validated assessment instruments, e.g., SGA or PG-SGA), body composition (body weight, BMI, muscle mass or lean body mass, fat mass) and/or objectively measured physical functioning (walk test, handgrip strength, physical or performance battery)	Quality of life, fatigue, feasibility, treatment tolerance or survival as only outcome measure
Study design	RCTs or pilot RCTs	Case series with <10 participants, qualitative studies, reviews, letters, editorials, notes
Setting	No restrictions	
Time frame	No restrictions	

Abbreviations: HNC: head and neck cancer; RT: radiotherapy; CT: chemotherapy; PEG: percutanous endoscopic gastrostomy; SGA: subjective global assessment; PG-SGA: patient generated-SGA; BMI: body mass index RCT: randomized clinical trial.

The details of the search process are shown in Figure 1. All identified records were screened for duplicates and irrelevant titles by the second author (JAS). Remaining abstracts were screened by three pairs of reviewers (LMO/LT, AB/TSS and GBS/JAS) and full-text papers were subsequently screened by the same pairs. Reasons for excluding abstracts and full-text papers were documented by the pairs. A third reviewer's opinion was called for in cases of disagreement regarding eligibility. Data concerning participant characteristics, content of the interventions, outcome measures, results and conclusions were extracted. Disagreement on final inclusion and exclusion were agreed by consensus by three of the authors (JAS, LMO, AB).

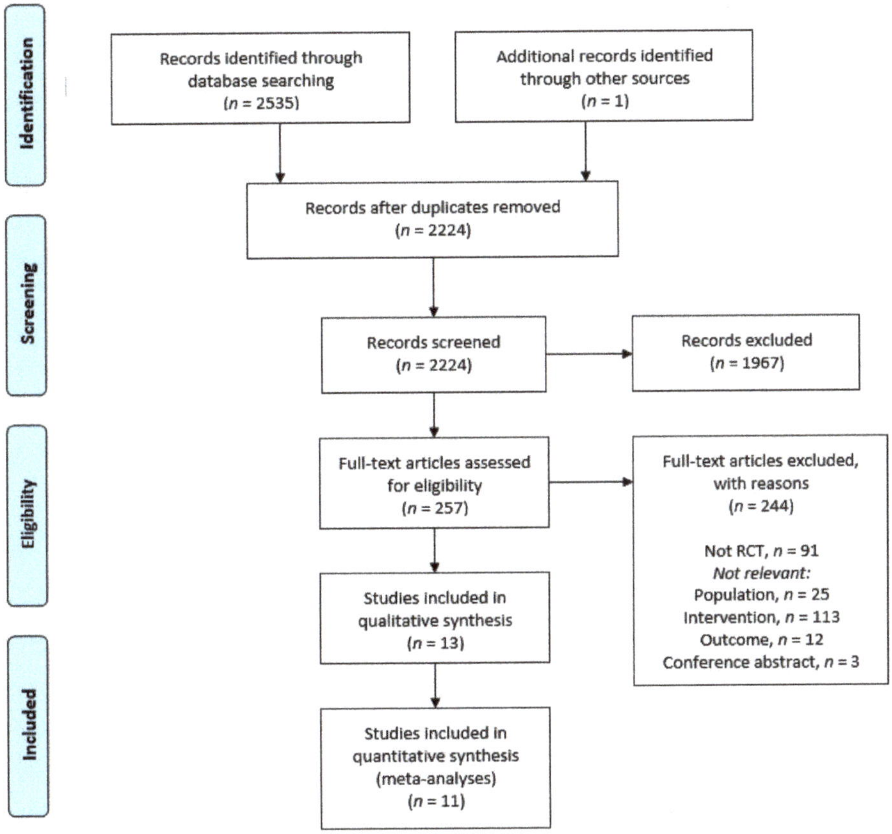

Figure 1. Preferred reporting items for systematic reviews and meta-analyses (PRISMA) flow diagram of reviewed and included studies.

2.2. Quality Assessment

The methodological quality of the included RCTs was assessed independently by two of the reviewers (AB, LMO), using the Physiotherapy Evidence Database (PEDro) scale [14]. The PEDro scale is developed for the Physiotherapy Evidence Database by the Centre for Evidence-Based Physiotherapy to evaluate the methodological quality of studies with physical exercise and therefore relevant for this review. The scale is also found to have acceptable reliability and validity when examining studies with other types of interventions [15]. The PEDro scale examines presence or absence of 11 quality measures, but only 10 are scored leaving the final score ranging from 0 to 10 points [14]. Criterion one relates to external validity (not calculated), criteria 2–9 assess internal validity, and criteria 10 and 11 verify whether the studies have enough statistical information for the results to be interpretable. A score between 8–10 is considered as high quality, 5–7 as moderate quality and 0–4 low quality.

2.3. Data Extraction and Statistical Analyses

The following data was extracted from the included studies by the first author (AB): authors, year, country, study design, patient group (sample size and disease), inclusion criteria, details of the interventions, adherence to the intervention, completion rate, outcomes and results. Adherence was defined to be reflected by measures of how well the patients complied with the intervention, e.g., energy intake in relation to calculated needs and number of exercise sessions completed. Whether the patients stayed in the trial to the end of study was registered as completion. Outcomes of interest were nutritional status, body composition and objectively measured physical function. Regarding nutritional status were use of validated assessment instruments (generic and disease specific) considered relevant as well as use of medical data known to reflect nutritional status. Regarding body composition, the following measures were considered relevant; absolute or change in body weight, body mass index (BMI), muscle mass, lean body mass or fat mass. For objectively measured physical function, the following measures were relevant; absolute or change in any physical fitness test such as a walk test, handgrip strength, physical or performance battery. When data were available and reliable scales were used, the studies were combined in a meta-analysis. We attempted to contact study authors to request values for any missing data and if this was not successful, we did not impute the data into the meta-analysis.

Publication bias was assessed by funnel plots and the *LFK* index and *Doi* plots [16] to detect and quantify asymmetry of study effects. LFK index values outside the interval between −1 and +1 are considered consistent with asymmetry (i.e., publication bias). Stata version 15 (Stata Corp, College Station, TX, USA) with the user-developed packages metan [17], metafunnel [18] metabias [19] and Doi plot and estimates the LFK index [20] were used for all the estimations.

3. Results

Search results are summarized in the Preferred Reporting Items for Systematic Reviews and Meta-Analyses (PRISMA) flow diagram (Figure 1). The database searches retrieved a total of 2535 records. One additional record was identified from a hand search of review paper references. After removal of duplicates, 2224 studies were left to screen. After screening title and abstracts, 1967 papers were excluded for not meeting the inclusion criteria, leaving 257 studies for full text review. After the full text review, 13 RCTs were included (Table 2), nine full scale RCTs and four pilot RCTs. Reasons for exclusions are listed in Figure 1. The included studies were conducted in Europe (n = 4), United States of America (n = 4), Asia (n = 3), Canada (n = 1), and Australia (n = 1). Publication year from 1984 [21] to 2019 [22].

3.1. Quality Assessment

Two of the full scale RCTs studies [23,26] were considered as high quality (eight to 10 points) (Table 3), four were considered of moderate quality (five to seven points) [22,27,29,30] and three as low quality [21,24,25]. Two of the pilot RCTs [28,31] were of high quality and two of moderate quality [9,32]. All the high and moderate quality studies clearly specified methods used and how the randomization was performed. Methodological uncertainties included blinding (minding that it is difficult to blind participants in both nutritional and exercise studies) and lack of intention to treat analysis [23,26,28]. For the three low quality studies, eligibility criteria were not specified [24] or unclear [25], it was uncertainties about random allocation to groups and similarity in outcome variables at baseline [21,24] as well as uncertainties regarding concealed allocation and intention-to-treat analysis [21,24,25].

Table 2. Characteristics of the included studies organized alphabetically.

Author, Year	Country	Study Design	Type of Intervention	Sample Size	Age in Years	Clinical Info Included Patients	Cancer Treatment
Capozzi, 2016 [10]	Canada	Pilot RCT	Exercise and nutrition	60 male 82%	Mean (SD) 56.1 (9.2)	Diagnosis (n): Larynx-hypopharynx (6), Nasopharynx (4), Oral (8), Oropharynx (29), Other (6), Unknown origin (7) Stage (n): I-III (11), IV (48) Histology: NA	RT (n = 16) CRT (n = 44)
Cereda, 2018 [23]	Italy	RCT	Nutrition	159 male 72%	Mean (SD) Counselling 63.8 (12.7); counselling + ONS 66.5 (14.5)	Diagnosis (n): Hypopharynx (13), Larynx (41), Naso- oropharynx (44), Oral (29), Other (32) Stage (n): 0-II (76), III (40), IV (34) Histology (n): Squamous (124), Lymphoma (20), other (15)	RT (n = 98) and CRT (n = 61)
Daly, 1984 [21]	USA	RCT	Nutrition	40 male 80%	Mean (SD), Intervention 53 (15) Control 55 (13)	Diagnosis (n): Nasopharynx (15), Other (25) Stage (n): II (2), III (9), IV (28) Histology: NA	All RT
Hearne, 1989 [24]	USA	RCT	Nutrition	31 male 77%	Mean (range), Intervention 52.1 (22–74) Control 56.1 (37–83)	Diagnosis (n): Nasopharynx (9), Other (22) Stage (n): III (11), IV (20) Histology: NA	All RT
Isenring, 2003 [25]	Australia	RCT	Nutrition	36 male 81%	Mean (SD), 63 (15)	Diagnosis: NA Stage: NA Histology: NA	All RT
Jiang, 2018 [26]	China	RCT	Nutrition	100 male 69%	Mean (SD), Intervention 46.7 (10.9)) Control 48.2 (11.1)	Diagnosis: Nasopharynx Stage: III (43), IVa-b (57) Histology: NA	All CRT

Table 2. *Cont.*

Author, Year	Country	Study Design	Type of Intervention	Sample Size	Age in Years	Clinical Info Included Patients	Cancer Treatment
Ravasco, 2005 [27]	Portugal	RCT	Nutrition	75 male 80%	Mean (SD) 60 (11)	Diagnosis (n NA): Larynx, Oropharynx, Nasopharynx, Tongue Stage (n): I-II (30), III-IV (45) Histology: NA	All RT (all pre-vious CT)
Rogers, 2013 [28]	USA	Pilot RCT	Exercise and nutrition	15 male 80%	Mean (SD) 60.5 (12.5)	Diagnosis (n): Nasopharynx, scalp and salivary glands (5), Other (10)Stage (n): I-II (7), III -IV (8) Histology: NA	RT (n = 11), CRT (n = 4)
Roussel, 2017 [29]	France	RCT	Nutrition	87 male 81%	Mean (SD) 60 (10)	Diagnosis (n): Hypopharynx (11), Larynx (19), Oral (9), Oropharynx (40), Nasopharynx (2), Sinus (2), Unknown origin (4) Stage (n): I-II (14), III (14), IV-V (59) Histology: NA	RT (n = 28), CRT (n = 59)
Sandmaell, 2017 [9]	Norway	Pilot RCT	Exercise and nutrition	41 male 61%	Mean (SD) 63.2 (9.3)	Diagnosis (n): Larynx (4), Nasal (1), Oral (5), Pharynx (20), Pharynx and larynx (1), Salivary glands (8), Unknown origin (2) Stage and histology: NA	RT (n = 24), CRT (n = 17)
Samuel, 2019 [22]	India	RCT	Exercise	148	Mean (SD) Intervention: 52.8 (9.7) Control: 52.8 (10.5)	Diagnosis (n): Larynx (28), Oropharynx (120) Stage (n): III (38), IVa (94), IVb (16) Histology: NA	All CRT
Samuel, 2013 [30]	India	RCT	Exercise	48 male 88%	Mean (SD) Intervention: 51.7 (10) Control: 52.5 (8.27)	Diagnosis: NA Stage: NA Histology: NA	All CRT
Zhao, 2016 [31]	USA	Pilot RCT	Exercise and nutrition	18 male 94%	Mean (SD) 57 (11)	Diagnosis (n): Larynx (1), Pharynx (15), Unknown origin (1) Stage (n): III (4), IV (7) Histology (n): NA	All CRT

Abbreviations: NA: not available; RT: radiotherapy; CRT: chemo/radiotherapy; RCT: randomized clinical trial.

Table 3. Methodological quality assessment: Randomized controlled trials on the effectiveness of exercise and/or nutrition interventions on nutritional status, physical function and quality of life in patients with head and neck cancer.

Study	Intervention Type	Criteria *											Total	Quality **
		1	2	3	4	5	6	7	8	9	10	11		
Randomized Controlled Trials														
Samuel, 2019 [22]	Exercise	+	+	+	+	–	–	–	+	–	+	+	7	Moderate
Samuel, 2013 [30]	Exercise	+	+	–	+	–	–	–	–	–	+	+	5	Moderate
Cereda, 2018 [23]	Nutrition	+	+	+	+	–	–	–	+	+	+	+	8	High
Jiang, 2018 [26]	Nutrition	+	+	+	+	–	–	–	+	+	+	+	8	High
Roussel, 2017 [29]	Nutrition	+	+	+	+	–	–	–	–	–	+	+	6	Moderate
Ravasco, 2005 [27]	Nutrition	?	+	+	?	–	–	–	+	+	+	+	6	Moderate
Isenring, 2003 [25]	Nutrition	–	+	?	+	–	–	?	+	–	?	+	4	Low
Hearne, 1989 [24]	Nutrition	–	?	?	?	–	–	–	–	–	+	+	2	Low
Daly, 1984 [21]	Nutrition	+	–	?	?	–	–	–	?	–	+	–	2	Low
Pilot and Feasibility Studies														
Sandmael, 2017 [9]	Exercise and nutrition	+	+	+	+	–	–	–	+	–	+	+	6	Moderate
Capozzi, 2016 [10]	Exercise and nutrition	+	+	+	+	–	–	+	–	+	+	–	7	Moderate
Zhao, 2016 [31]	Exercise and nutrition	+	+	+	+	+	+	+	+	+	+	+	10	High
Rogers, 2013 [28]	Exercise and nutrition	+	+	+	+	–	–	–	+	+	+	+	8	High

* The criteria addressed the following issues: 1 eligibility criteria specified; 2 randomly allocated to groups; 3 allocation concealment; 4 groups similar at baseline; 5 blinding of all subjects; 6 blinding of caregivers; 7 blinded outcome assessment; 8 measures obtained from least 85% of subjects; 9 intention-to-treat analysis; 10 between-group statistics; 11 measure of variability. + = yes, - = no and ? = unclear. Points were awarded only when a criterion was clearly satisfied. Criterion 1 is not scored. Each other criterion was given equal weight (i.e., 1 point) for a maximum sum score of 10. ** High quality: 8–10, moderate: 5–7, low: 0–4.

3.2. *Study Characteristics*

Four pilot RCTs (Table 2) investigated effects of interventions combining nutrition and physical exercise [9,28,31,32] with sample sizes between 15 [28] and 60 participants [32]. Seven studies investigated the effects of nutritional interventions only [21,23–27,29] with sample sizes varying from 31 [24] to 159 participants [23] and study duration ranging from six weeks during treatment [26] up to six months due to follow up after the intervention period [21] (Tables 2 and 4). One study had three arms [27], i.e., one group received individualized dietary counselling, one ONS and the last group was advised to eat ad libitum. Two studies with sample sizes of 48 and 148 patients investigated exercise interventions during RT with follow up seven and four weeks after end of RT, respectively [22,30] (Table 4).

Table 4. Description of intervention, length of follow-up, adherence to intervention and completion rate, organized according to design, year of publication and intervention.

Study	Intervention Type	Description of Intervention	Length of Intervention	Intervention Adherence	Completion Rate
Randomized Controlled Trials					
Samuel, 2019 [22]	Exercise	Intervention: Brisk walking for 15–20 min and resistance training for major muscles of upper and lower limb, 2 sets and 8–15 repetitions. Exercise sessions monitored at the hospital, five days a week followed by a monitored home-based program. Control: Physical exercise recommendation, 10 min walks during the day five days a week.	Seven weeks during RT at the hospital followed by four weeks at home	NA	120/148 Lost: Intervention 16 Control 12
Samuel, 2013 [30]	Exercise	Intervention: Brisk walking for 15–20 min at perceived exertion rate between 3–5/10, five days a week. Individually tailored program for major muscle groups of upper and lower limbs 2–3 sets and 8–10 repetitions. Exercise sessions five days a week. Control: No scheduled exercise sessions but advised to remain as physically active as possible.	Intervention during RT, 6 weeks	NA	43/48 Lost: Intervention 4 Control 1
Cereda, 2018 [23]	Nutrition	Intervention: Nutritional counseling based on estimated protein–calorie requirement (1.2 g/kg of actual body weight), personal eating patterns and preferences, chewing and swallowing abilities. Addition of 1–2 bottles/day of n–3 polyunsaturated fatty acids–enriched ONS. Follow–up during RT: once a week for 6 weeks. After RT: one month and three months Control: Nutritional counseling as described above. No n–3 ONS but for ethical reasons ONS were prescribed when food intake was too low (< 60% of estimated requirements for two consecutive weeks). EN or PN was started if intake was too low for two consecutive weeks despite the use of ONS.	During RT and 3 months follow–up	NA, but protein intake (g protein/kg/day) described: End of RT: Intervention 1.0 vs. control 0.87 1 month after RT: Intervention 1.16 vs. control 0.97 3 months after RT: Intervention 1.12 vs. control 0.96	112/159 Lost: Intervention 22 Control 25

Table 4. *Cont.*

Study	Intervention Type	Description of Intervention	Length of Intervention	Intervention Adherence	Completion Rate
Randomized Controlled Trials					
Jiang, 2018 [26]	Nutrition	Intervention: ONS 100g/day (402 kcal, 18 g protein) Control: No ONS General dietary advices in both groups every week PN with glucose if intake was severely compromised	During CRT	Consumed 52.1 g (29.4g)/day	91/100 Lost: Intervention 5 Control 4
Roussel, 2017 [29]	Nutrition	Intervention: Six individualized counselling meetings with a dietitian at home (two during RT and four at the end of RT). One meeting 2 months after end of RT. Energy and protein requirements individually evaluated and nutritional adjustments obtained with regular foods, ONS or EN if necessary. Education for self–monitoring weight, adapting intake and modifying food textures. Control: As described above but only two outpatient consultations with a dietitian during RT. Recalls if needed.	During RT, 3 months follow–up	NA but energy intake (kcal/kg/day) described: 1 month after RT: Intervention 34 vs. control 33 3 months after RT: Intervention 35 vs. control 31	87/117 Lost: Intervention 16 Control 14
Ravasco, 2005 [27]	Nutrition	Group 1 (n = 25): individualized counselling with regular foods Group 2 (n = 25): usual diet plus ONS (2 × 200 mL containing 20 g protein and 200 kcal per day) Group 3 (n = 25): intake ad libitum Nutritional goal for group 1 and 2 was achievement of individually calculated energy and protein requirements	Intervention during RT, 3 months follow–up	NA, but nutritional intake was primary endpoint and reported Baseline: intake similar in all groups End of RT: group 1 increase of 521 kcal/day, $p = 0.002$ ONS increase of 322 kcal/day, $p = 0.05$ Ad lib decrease of 400 kcal/day, $p \leq 0.01$ Between–group finding, $p = 0.005$ 3 months: group 1 maintained energy intake, other groups decreased, $p = 0.001$	All completed

Table 4. *Cont.*

Randomized Controlled Trials					
Study	**Intervention Type**	**Description of Intervention**	**Length of Intervention**	**Intervention Adherence**	**Completion Rate**
Isenring, 2003 [25]	Nutrition	Intervention: Individualized counselling by using a standard protocol (American Dietetic Association Medical Nutrition Therapy Head and Neck). ONS were provided when appropriate Control: Regular care, general advice by the nursing staff with samples of ONS if felt necessary.	Intervention during RT, 3 months follow−up	NA	32/36 Lost: Intervention 1 Control 3
Hearne, 1985 [24]	Nutrition	Intervention: Intensive nasogastric feeding during RT Control: Oral intake and dietary counselling Goal for intervention in both groups: 40 kcal/kg per day and 1g protein/kg per day	Intervention during RT, 1 month follow−up	Intervention: Two of 14 (14%) refused tube feeding and converted to control Control: Two of 12 (16%) converted to intervention due to weight loss Energy intake during RT (kcal/kg): Intervention 35–42 vs. control 15–34 No p−values given.	26/31 Lost: Intervention 4 Control 1
Daly, 1984 [21]	Nutrition	Intervention: EN Control: Oral intake and dietary counselling Goal in both groups: 40 kcal/kg per day and 1–1.5 g protein/kg per day. If weight gain did not occur after each week +5 kcal/kg per day. Both groups received enteral support throughout RT (approximately 8 weeks) and for several additional weeks until reaction to radiation subsided	Intervention during and up to 6 months follow−up	Intervention: Two of 22 (9%) converted to control due to non−compliance during the first week of RT Control: Two of 15 (11%) converted to tube feeding due to weight loss during the two first weeks of RT Energy intake (kcal/kg): Tube fed 39 vs. orally 30, $p < 0.00$	35/38 Lost: NA

Table 4. *Cont.*

Pilot and Feasibility Studies					
Study	**Intervention Type**	**Description of Intervention**	**Length of Intervention**	**Intervention Adherence**	**Completion Rate**
Sandmæl, 2017 [9]	Exercise and nutrition	Group 1: During treatment: Resistance exercises: 2 lower body– and 2 upper body, 3–4 sets, 6 to 12 repetitions, monitored by a physiotherapist at the hospital twice a week á 30 min (total 12 sessions). Recommended 150 min of moderate intensity exercise per week in addition. After the training sessions one bottle ONS. Recommended to take 1–2 ONS each day. Group 2: During treatment: Recommended to follow physical exercise guidelines for cancer patients. 2–4 weeks after end of RT: 3 weeks stay at rehabilitation centre. Resistance exercises: 3 sessions of 45 min of involving 3 upper body and 3 lower body exercises. 3–4 sets and 6 to 12 repetitions plus two voluntary sessions each week involving a combination of strength, aerobic and balance exercises with low intensity. Dietary counselling once a week in small groups and use of ONS.	Intervention during RT for group 1 and intervention after RT for group 2 Intervention initiated *during the first week* of radiotherapy lasting 6 weeks.	Adherence rates (%): Interv during RT, exercise 81 and ONS 57 After RT, exercise 94 and ONS 76	29/41 Lost: Intervention 2 Control 10
Zhao, 2016 [31]	Exercise and nutrition	Group 1: Intervention based on guidelines for patients with cancer (American College of Sports Medicine); strengthening, cardiovascular fitness and physical exercise. Exercise during the 7 weeks CRT at a clinical research center supervised by a trainer. Up to 3 sessions per week, lasting up to 1 h including warmup, cool down, and rest periods. Resistance exercises included chest press, wall push up, military press, side arm raises, biceps curl, shoulder shrugs, and calf raises. Duration and intensity were customized to the individual, goal three 8 to 12 repetition sets. Aerobic exercise was defined as walking with a pedometer and a goal to maintain step count based on the mean step count of the previous training week. Post CRT (weeks 8 to 14), integration of exercise activities into own lifestyle. Weekly telephone calls from the trainer. Before CRT counselling by a dietician, repeated in case of decrease in BMI greater than a 5% to 10%. Group 2: Standard treatment, dietary counselling and active nutritional surveillance during RT, neither encouraged nor discouraged to exercise.	Intervention for 7 weeks 7 weeks follow up	Exercise adherence rate 72%, Completed 15.2 of 21 sessions.	17/20 Lost: Intervention 1 Control 2

Table 4. *Cont.*

Pilot and Feasibility Studies					
Study	**Intervention Type**	**Description of Intervention**	**Length of Intervention**	**Intervention Adherence**	**Completion Rate**
Rogers, 2013 [28]	Exercise and nutrition	Group 1: Nutritional counseling and 12 weeks resistance exercise. Exercise during treatment; one hour supervised sessions twice weekly at a training facility at the hospital. Six weeks of twice weekly home–based sessions supported with telephone counseling, written materials, and DVD. Up to 10 repetitions of 9 different exercises using a resistance band for major muscle groups (chest press, leg extension, lateral row, reverse curl, triceps using wall push–up/triceps kickback, heel raise, 2–arm front raise, hamstring curl, and arm curl). Intensity: light, moderate and heavy resistance bands were used. Group 2: Nutritional counseling provided by registered dietitian according to standard counseling appropriate for head and neck cancer during radiotherapy	12 weeks intervention	Exercise adherence: 6 weeks: 83% 6–12 weeks: 62% Both groups Face to face nutritional counselling (6 weeks): 96% completed Telephone counselling (6–12 weeks): 77% completed	13/15 Lost: Intervention 2 Control 0
Capozzi, 2016 [10]	Lifestyle interventions including exercise and nutrition	Group 1: 12 weeks lifestyle intervention during RT and Group 2: same intervention immediately after completion of RT. Components of intervention: physician referral and clinical support; health education; behavioral change support; individual exercise program (home exercises twice a week); group–based exercise (2 exercise sessions weekly) Exercise program: progressive resistance training with 2 sets of 8 repetitions at 8 to 10 repetitions maximum for 10 exercises targeting major muscle groups. In addition to exercise sessions participants were required to attend 6 education sessions biweekly	Immediate intervention during RT for group 1 and delayed intervention after RT for group 2 Total 24 weeks	NA	36/60 Lost: Group1: 15 Group 2: 9

Abbreviations: NA: not available; RT: radiotherapy; ONS; oral nutritional supplements; EN: enteral nutrition; PN: parenteral nutrition; CRT: chemo/radiotherapy; RCT: randomized clinical trial.

3.3. *Effects on Nutritional Status, Body Composition and Physical Function*

Outcomes and effects of the interventions are summarized in the Supplementary table. Nutritional status was measured in three studies [26,27,32]. No statistically significant difference between intervention and control group were found. Just two of the studies [26,32] presented group data and therefor a quantitative analysis of effects on nutritional status was not meaningful.

Nine studies were included in the quantitative synthesis of effects on body composition (Figure 2a) [9,23–26,28,29,31,32]. Absolute weight or weight change were used as outcome variable for body composition in all studies except for three [28,31,32] were change in BMI was used. In the fixed-effect meta-analysis on body composition, it was a significant difference between intervention and control group for the studies using only nutrition as intervention (SMD 0.42 (95CI 0.23 – 0.62), $p < 0.001$), but not for the trials combining nutrition and physical exercise (SMD 0.21 (95CI −0.16 – 0.58), $p = 0.259$). Still, the estimated difference using all the included trials was highly significant (SMD 0.38 (95CI 0.20 – 0.55), $p < 0.001$) with a better outcome for the intervention group. The heterogeneity was low with an overall I^2 statistics of 0% and a non-significant Cochran's Q test (p-value = 0.463). Assessing the corresponding funnel plot (Supplementary Figure S1) and the Egger's test for small-study effects ($p = 0.947$) as well as the DOI plot and the LFK index (−0.99) (Figure 2b), no publication bias in the studies were detect.

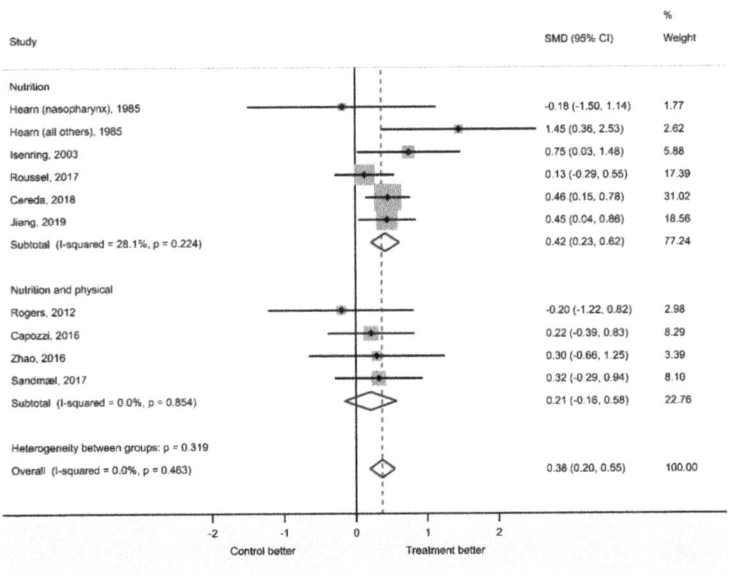

(a)

Figure 2. *Cont.*

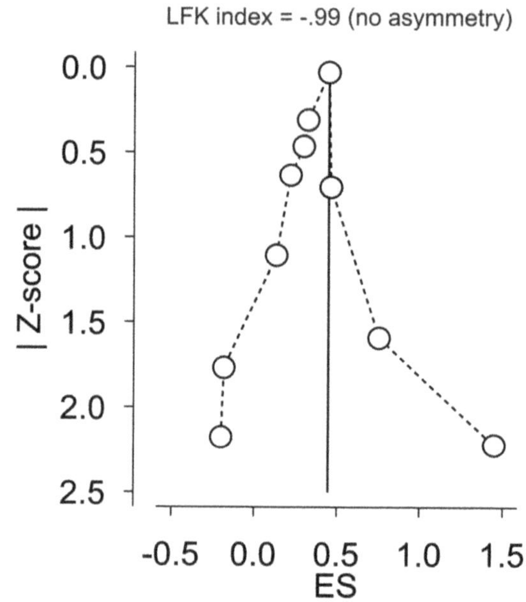

(**b**)

Figure 2. (**a**) The effects of nutritional and exercise interventions on body composition; (**b**) the symmetry of body composition results presented in Doi plot. ES, effect size.

Five studies were included in the quantitative synthesis of effects on physical function (Figure 3a) [22,28,30–32], of which three studies combined nutrition and physical exercise [28,31,32]. The six minutes-walk test was used as outcome measure except for Capozzi et al. [32] and Rogers et al. [28] where handgrip strength was used. In the fixed-effect meta-analysis on physical function, it was a highly significant difference between intervention and control group (SMD 0.78 (95CI 0.51–1.04), $p < 0.001$) with a better outcome for the intervention group. The heterogeneity was higher than in the trials on body mass with an overall I^2 statistics of 50.1%, but a non-significant Cochran's Q test (p-value = 0.091). Assessing the corresponding funnel plot (Supplementary Figure S2) and the Egger's test for small-study effects ($p = 0.896$) as well as the DOI plot and the LFK index (−0.73) (Figure 3b), no publication bias in the studies were detected.

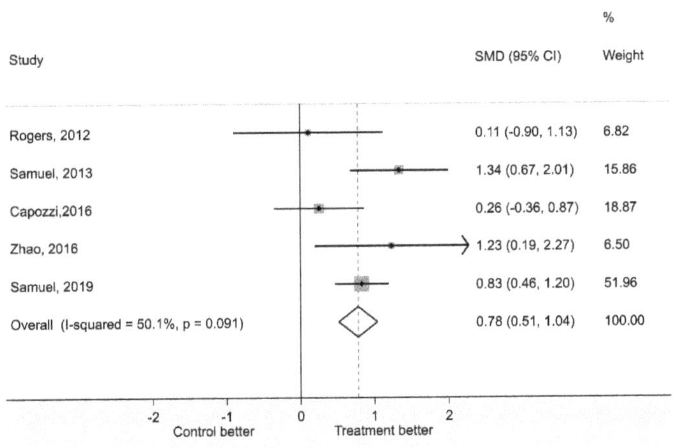

(a)

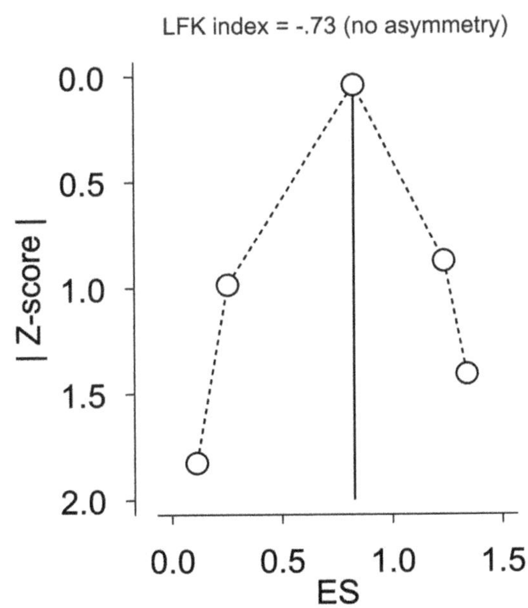

(b)

Figure 3. (**a**)Effects of nutritional and exercise interventions on physical function; (**b**) the symmetry of physical function results presented in Doi plot.

3.4. The Content of the Interventions

A detailed description of the content in the interventions is presented in Table 4.

3.4.1. Nutrition

The most frequent nutritional intervention (six of 11 studies) was individualized dietary counselling based on regular food with or without ONS aiming to meet estimated individual needs for energy and protein [23,25,27–29,31]. In three of the studies [23,25,29], dietary counselling was considered as standard care and therefor applied in the control group but with less monitoring and feedback than in the intervention group. In one study, participants in both groups received dietary counselling by a dietitian before initiation of CRT, and then again at discretion of the attending physician, if the participants experienced a decrease in BMI of 5 to 10% [31]. Two studies intervened with ONS only and the patients were encouraged to take 1-2 bottles each day [9,26]. Nasogastric tube feeding was the intervention in two studies whereas the control group received dietary counselling [21,24]. For both tube feeding and counselling, the goal was to meet estimated energy (40 kcal/kg/day) and protein (1 to 1.5 g protein/kg/day) needs. For the last study [32] the exact content of the nutritional intervention was not specified, but it was reported that a group based dietary counselling was given by a dietitian as part of a 12-week lifestyle program.

3.4.2. Physical Exercise

In four of the six studies that included physical exercise, a combination of resistance and aerobic exercises was applied [9,22,30,31]. Two studies intervened with resistance exercises only [28,32]. The resistance exercises covered the major muscle groups and were monitored and supervised by a trainer or physiotherapist 2–5 times a week during RT in five studies [9,22,28,31,32]. In the last study [30] the patients received an individualized and structured exercise program, and their family members were asked to motivate the patients to do the exercises. In all studies, the patients were encouraged to proceed with the resistance exercises at home after RT and a weekly follow up telephone was applied in three studies [22,28,31]. The applied aerobic exercise included brisk walking for 15–20 min five days a week [22,30], multiple short duration continuous walking to achieve a total walking time of 30 min a day [31] and 150 min moderate intensity aerobic exercise per week [9].

3.5. *Adherence to the Intervention and Completion Rate*

Adherence to the interventions and completion rates are presented in Table 4.

3.5.1. Adherence

Four studies [25,30–32] did not present data regarding nutritional interventions adherence. Three studies evaluated adherence in relation to how well the patients met their energy and protein needs [23,27,29]. In two of these studies [23,29] intake in accordance with estimated needs was reported during the intervention and follow-up both in the intervention and control group. Ravasco et al. [27] found that the group receiving dietary counselling had higher energy intake and thereby better adherence than the group using ONS and the ad lib group at the end of RT and at three months follow-up. In the two studies using ONS, adherence was evaluated as number of ingested ONS in relation to planned amount. One study reported an adherence rate of 57% during treatment and 76% after treatment [9] while the other study reported that about 52% of the provided ONS were consumed [26]. In the two studies investigating tube feeding, 9% [24] and 14% [21] refused the intervention and were converted to the control group. In the same studies, were two patients converted from the control to the intervention group in each study due to weight loss during the first week of RT.

Four of six studies reported data on adherence to the exercise intervention [9,28,31,32]. Rogers et al. [28] reported that 83% of the planned exercise sessions were completed at 6 weeks and 62% in the period between week 6 and 12. In Zhao et al., [31] the overall adherence to the exercise program was 72% (15.2 out of maximum 21 sessions). Two studies had a similar design with an intervention during RT and another group receiving a delayed intervention, i.e., after completion of RT [9,32]. In the study of Capozzi et al. [32] the weekly attendance to the supervised exercise program was 45% during cancer treatment and 61% after. In Sandmael et al. [9] overall adherence to strength and aerobic exercise was 81% and 94%, respectively.

3.5.2. Completion Rates

All studies reported data on completion rate. Ravasco et al. [27] reported that all patients completed the study. For the other studies with nutritional intervention the completion rate varied between 70% [23] and 92% [21]. For most nutritional studies patients lost for follow up was similar in the intervention and control group except for the study of Isenring et al. [25] (7% in the intervention and 14% in the control group) and Hearne et al. [24] (22% in the intervention and 8% in the control group).

The exercise only studies reported 81% [22] and 86% [30] completion and similar number of patients lost for follow-up in the intervention and control group. In the feasibility studies investigating a combination of nutrition and exercise, the completion rate varied between 60% [32] and 87% [28].

4. Discussion

This systematic review and meta-analyses show that nutrition and physical exercise interventions have a positive effect on body composition and physical function for patients with HNCs undergoing RT (+/- concomitant CT) with a curative intent. The nutritional interventions were mainly individualized dietary counselling aiming to meet estimated energy and protein needs and use of ONS in case of inadequate energy intake. The physical exercise was typically supervised with a combination of strength and aerobic exercises used, performed two to five times a week. In case of nutritional interventions, the adherence to dietary advices after counseling was reported good, but it was measured in just half of the studies. When ONS were used, about half of the patients did not consume the recommended amount. The adherence to exercise varied between 45% and 83% and completion rates between 60% and 80%. The lowest adherence and completion rate were reported for interventions combining nutrition and physical exercise.

4.1. Strengths and Limitations

This is, to our knowledge, the first systematic review seeking to examine the effects of both nutrition and physical exercise in patients with HNCs undergoing RT. A major strength of this review is the authors' attempt to identify all relevant studies by using a comprehensive search strategy in multiple databases lead by a research Liberian as well as methodological strictness performing the systematic review and meta-analyses. All authors participated in the process which also included hand-search of review paper references to identify additional studies that may have been lost in the initial search.

Based on available guidelines, it was expected that interventions combining nutrition and physical exercise would have a better effect on nutritional status, body composition and physical function than nutrition and physical exercise alone [5]. However, only four studies [9,28,31,32] with combined interventions were identified and included in the meta-analysis to explore the effects on body composition. A major limitation is that all four studies were pilot/feasibility, i.e., not powered to detect statistically significance difference between the groups. Another limitation was that no relevant measures regarding the other outcomes of interest, nutritional status and physical function, were provided. Based on this, it is not possible to draw any meaningful conclusion regarding effects of combined nutrition and physical exercise interventions in patients with HNCs undergoing RT.

Several factors, largely reflecting limitations in the included studies, may have influenced the results showing effect on body composition and physical function of the interventions. Nine of 13 studies were of poor or modest methodological quality mainly due to uncertainties about baseline assessments of outcome variables, heterogeneity in anti-cancer treatment and random allocation to groups. In addition, uncertainties regarding intention-to-treat analysis were seen in six of the studies. The lowest methodological quality was seen in the oldest studies [21,24,25,27], all investigating effects of nutritional interventions. One of these studies used only within group and not between-group statistical comparisons analyzing the outcomes of interest for this review [27] which make the results more or less useless in a randomized design where the aim is to compare two or more groups.

The specific interventions given in the included studies were heterogenous and in many studies poorly described. Even if individualized dietary counselling and combinations of strength and aerobic exercises were the most common interventions, it was a variation in the delivery that may have affected the results. In one of the studies [31] the nutrition intervention was delivered as part of a comprehensive lifestyle program. Thus, participants were receiving concomitant additional lifestyle interventions such as clinical support and health education which may have had a synergistic effect on the outcomes. Additionally, parts of the lifestyle program were also used in the control group potentially contributing to an equalization of possible effects [33].

The measurements of outcomes of interest for this review were highly heterogeneous. In the nutrition field it is an acknowledged problem that high quality indicators to demonstrate the effect of nutritional interventions are lacking [34]. Changes in weight and BMI have long been regarded as practical indicators of changes in nutritional status and body composition [35]. Although of value, these measurements do not captured changes in muscle mass which is associated with several negative outcomes specifically in cancer patients [35,36]. The use of weight and BMI as measures of body composition may have confounded the effects of the nutrition intervention but may even more the interventions with physical exercise since they are expected to have a direct effect on muscle mass. The exercise studies were also heterogenous regarding the measurements of physical function (three used six-minute walk test and two used hand grip test) and one study [9] did not include an objective physical functioning at all. The most used six-minute walk test mainly measure walking ability and endurance and may not catch up changes in muscular strength, muscle mass and muscle waste [37,38]. Thus, future full scaled studies including both nutrition and physical exercise are warranted. The future studies should more carefully choose an appropriate and specific method to measure body composition and physical function according to the intervention given.

4.2. Nutritional Interventions

The result showing that nutritional interventions alone have a positive effect on body composition is in line with the results from a former study reviewing effects of nutritional interventions on nutritional status, quality of life and mortality in patients with HNCs receiving RT [7]. The authors concluded that individualized dietary counselling based on regular food with or without ONS has a beneficial effect on energy and protein intake and nutritional status when comparing with standard nutritional care. Additionally, they found that ONS alone only showed short-term effects on energy and protein intake and inconsistent effects on nutritional status and tube feeding versus ONS showed no beneficial effects.

The current nutrition guidelines for patients with HNCs recommends individualized dietary counselling in combination with ONS and/or initiation of tube feeding when oral intake is inadequate [5]. In the present review, this approach was used in five of 11 studies with nutritional intervention [23,25,26,29,31] while one used only dietary counselling [28] and a pilot study used only ONS [9]. Dietary counselling is considered the best approach to promote adherence to dietary advices [39,40] since it allows an individual tailoring of the diet to personal needs and desires [40,41]. An indication of this was also found in one of the selected studies, designed to compare effect on dietary intake after dietary counselling, use of ONS and eating ad libitum [27]. It was concluded that dietary counselling was the only intervention that improved dietary intake and had a positive effect on nutritional status. However, a more recent study found that HNCs patients receiving counselling in combination with ONS had a higher intake of micronutrients and preserved weight better than patients not using ONS [42], supporting the current guidelines recommending addition of ONS when oral intake is inadequate [5].

Two older studies of low methodological quality used tube feeding from start of RT [21,24]. A recent review did not show that prophylactic tube feeding in patients with HNCs is more beneficial than ONS regarding nutritional status and body composition [7]. However, after individual considerations tube feeding may be regarded beneficial, but since some patients may consider it burdensome, it is important to explore the patient's wishes and preferences before initiated [43,44].

Unfortunately, there was little information about adherence to dietary counselling, as it was reported in only three of six studies [23,27,29]. All studies reported dietary intake in accordance with estimated needs indicating high adherence. This supports the assumption that individualized dietary counselling promote adherence to dietary advices [40]. The study of Ravasco et al. [27] also found that counselling resulted in a higher intake of macronutrients than just using ONS. This is in line with the findings from the two included studies using ONS as intervention [9,26], both showing low adherence (57% and 52%, respectively). In a qualitative study from our group the respondents with HNCs expressed that ONS only made sense during the initial weeks of radiotherapy, and that after this it got unbearable to ingest them due to side effects from RT [45]. These respondents also indicated that being exposed to the side effects of radiotherapy was experienced as quite different from just hearing and reading about them. This finding may have consequences for when nutritional interventions should be delivered. It is possibly not necessary to use intensive dietary counselling from start of RT, but instead use nutritional surveillance systematically and provide of dietary counselling when the patients developed eating problems as recommended in a study [46].

4.3. Exercise Interventions

According the recently published guidelines for physical exercise in cancer patients, there is relative strong evidence for prescribing physical exercise for the effects on physical function for cancer patients [47]. However, it should be noted that these guidelines are based on data from self-reported physical function (using different self-reported questionnaires) and not results for objective physical function being used as the outcome in this meta-analysis. Regarding data from objective measures, the evidence base on this outcome remains immature and more challenging to aggregate due to the variation and limitations of assessment techniques. Therefore, the results from our meta-analysis needs to be regarded with caution and more studies are warranted to conclude more firmly.

Four of six potential studies reported adherence to the exercise intervention, and reported adherence was in general high, but ranging from 45–83%. However, the reporting of exercise was different between the studies, making it challenging to compare. Sandmæl et al. [9] reported a high adherence rate of 81% for the entire period during treatment, while Rogers et al. [28], divided the adherence rate in the period between 0–6 weeks (83%) and 6–12 weeks (62%) showing a decline in adherence in the six last weeks of treatment as the patients get more complaints. The reporting of adherence in exercise studies has until recently been suboptimal in most studies and in the future, greater demands should be made concerning reporting of adherence [37].

Supervised exercise appears to be more effective than unsupervised or home-based interventions [47,48]. In line with these recommendations, all the included studies in this meta-analysis used supervised exercise.

5. Conclusions

This meta-analysis found significantly positive effects of interventions with nutrition alone and physical exercise alone in body composition and objective physical function in favor of the treatment groups. However, the included studies were highly heterogenic both regarding measurement methods and the content of the interventions which may have affected the result of the meta-analysis. Due to the pilot and feasibility design of the studies combining physical exercise and nutrition, no conclusions can be drawn concerning the effects from these studies. Future full-scaled RCTs combining nutrition and physical exercise is warranted.

Supplementary Materials: The following are available online at http://www.mdpi.com/2072-6643/12/11/3233/s1, Figure S1: The symmetry of body composition results presented in Funnel plot, Figure S2: The symmetry of physical function results presented in Funnel plot, Table S1: Outcomes and effects of nutritional and exercise interventions, organized according to study type, year of publication and intervention.

Author Contributions: Conceptualization, A.B., L.M.O., J.A.S., G.B.S., L.T., T.R.B. and T.S.S.; methodology, A.B., L.M.O., J.A.S., A.H.P., G.B.S., L.T., T.R.B. and T.S.S.; formal analysis, A.B., A.H.P. and L.M.O.; writing—original draft preparation, A.B. and L.M.O.; writing—review and editing, A.B., L.M.O., J.A.S., A.H.P., G.B.S., L.T., T.R.B. and T.S.S. All authors have read and agreed to the published version of the manuscript.

Funding: The present manuscript is conducted as part of a PhD thesis funded by the Norwegian ExtraFoundation for Health and Rehabilitation.

Acknowledgments: Thanks to Ingrid I. Riphagen, adviser at the DMF faculty administration, Norwegian University of Science and Technology (NTNU), Trondheim, for great help with the electronic searches and valuable feedback on the protocol.

Conflicts of Interest: The authors have no conflicts of interest associated with the present manuscript.

Appendix A Search Strategy

Appendix A.1 PubMed—3 June 2019

#1. "Head and Neck Neoplasms" [Mesh] AND ("Diet Therapy" [Mesh] OR "diet therapy" [Subheading] OR "Dietary Supplements"[Mesh] OR "Exercise"[Mesh] OR "Exercise Movement Techniques"[Mesh] OR "Exercise Therapy"[Mesh])

#2. ((head[ti] OR neck[ti]) AND (cancer[ti] OR tumor[ti] OR tumour[ti] OR carcinoma*[ti])) AND (exercis*[ti] OR diet[ti] OR diets[ti] OR dietary OR nutrition*[ti] OR training[ti] OR physical activity[ti] OR rehabilitation[ti] OR life style[ti]) NOT medline[sb]

#3. #1 OR #2 > 552 hits > EndNote PubMed in label field

Appendix A.2 Embase—1974 to 3 June 2019

1. "head and neck cancer"/dm, rt, rh, si, th [Disease Management, Radiotherapy, Rehabilitation, Side Effect, Therapy]
2. (diet therapy/or dietary intake/or exp exercise/or exp kinesiotherapy/or nutritional counseling/or nutritional support/or diet supplementation/or nutrition/)
3. 1 and 2 > 350 hits > EndNote Embase in label field

Appendix A.3 Cochrane Library—CDSR issue 6/12, June 2019, DARE issue 2/4, April 2015, CENTRAL issue 5/12, May 2016

#1. (head or neck) and (cancer or carcinom* or tumor* or tumour*): ti,ab,kw

#2. (exercise or training or diet or diets or dietary or nutrition or rehabilitation or "life style" or "physical activity"): ti,ab,kw

#3. #1 and #2 > 379 hits (40 CDSR/7 DARE/343 CENTRAL) > EndNote CDSR/DARE/CENTRAL in label field

Appendix A.4 CINAHL June 2019

S1 TI ((head OR neck) AND (cancer OR tumor OR tumour OR carcinoma*))

S2 AB ((head OR neck) AND (cancer OR tumor OR tumour OR carcinoma*))

S3 TI (exercis* OR diet OR diets OR nutrition* OR training OR rehabilition OR "physical activity" OR lifestyle OR "life style")

S4 AB (exercis* OR diet OR diets OR nutrition* OR training OR rehabilition OR "physical activity" OR lifestyle OR "life style")

S5 TI (therap* OR treatment* OR intervention* OR management* OR radiotherap* OR chemotherap* OR chemoradiotherap*)

S6 AB (therap* OR treatment* OR intervention* OR management* OR radiotherap* OR chemotherap* OR chemoradiotherap*)

S7 (s1 OR s2) AND (s3 OR s4) AND (s5 or s6) > 389 hits > EndNote CINAHL in label field

References

1. Forouzanfar, M.H.; Foreman, K.J.; Delossantos, A.M.; Lozano, R.; Lopez, A.D.; Murray, C.J.; Naghavi, M. Breast and cervical cancer in 187 countries between 1980 and 2010: A systematic analysis. *Lancet* **2011**, *378*, 1461–1484. [CrossRef]
2. Ratko, T.A.; Douglas, G.W.; de Souza, J.A.; Belinson, S.E.; Aronson, N. Radiotherapy treatments for head and neck cancer update. In *AHRQ Comparative Effectiveness Reviews*; Comparative Effectiveness Review No. 144; Agency for Healthcare Research and Quality (US): Rockville, MD, USA, 2014.
3. Kubrak, C.; Olson, K.; Jha, N.; Jensen, L.; McCargar, L.; Seikaly, H.; Harris, J.; Scrimger, R.; Parliament, M.; Baracos, V.E. Nutrition impact symptoms: Key determinants of reduced dietary intake, weight loss, and reduced functional capacity of patients with head and neck cancer before treatment. *Head Neck* **2010**, *32*, 290–300. [CrossRef]
4. Kilgour, R.D.; Vigano, A.; Trutschnigg, B.; Hornby, L.; Lucar, E.; Bacon, S.L.; Morais, J.A. Cancer-related fatigue: The impact of skeletal muscle mass and strength in patients with advanced cancer. *J. Cachexia Sarcopenia Muscle* **2010**, *1*, 177–185. [CrossRef]
5. Arends, J.; Bachmann, P.; Baracos, V.; Barthelemy, N.; Bertz, H.; Bozzetti, F.; Fearon, K.; Hütterer, E.; Isenring, E.; Kaasa, S. ESPEN guidelines on nutrition in cancer patients. *Clin. Nutr.* **2017**, *36*, 11–48. [CrossRef]
6. Findlay, M.; Bauer, J.; Brown, T.; Davidson, W.; Hill, J.; Isenring, E.; Talwar, B.; Bell, K.; Kiss, N.; Kurmis, R. Evidence based practice guidelines for the nutritional management of patients with head and neck cancer. *Clin. Guidel. Netw.* **2011**.
7. Langius, J.A.; Zandbergen, M.C.; Eerenstein, S.E.; van Tulder, M.W.; Leemans, C.R.; Kramer, M.H.; Weijs, P.J. Effect of nutritional interventions on nutritional status, quality of life and mortality in patients with head and neck cancer receiving (chemo) radiotherapy: A systematic review. *Clin. Nutr.* **2013**, *32*, 671–678. [CrossRef]
8. Garg, S.; Yoo, J.; Winquist, E. Nutritional support for head and neck cancer patients receiving radiotherapy: A systematic review. *Supportive Care Cancer* **2010**, *18*, 667–677. [CrossRef]
9. Sandmael, J.A.; Bye, A.; Solheim, T.S.; Stene, G.B.; Thorsen, L.; Kaasa, S.; Lund, J.A.; Oldervoll, L.M. Feasibility and preliminary effects of resistance training and nutritional supplements during versus after radiotherapy in patients with head and neck cancer: A pilot randomized trial. *Cancer* **2017**, *123*, 4440–4448. [CrossRef]
10. Capozzi, L.C.; Nishimura, K.C.; McNeely, M.L.; Lau, H.; Culos-Reed, S.N. The impact of physical activity on health-related fitness and quality of life for patients with head and neck cancer: A systematic review. *Br. J. Sports Med.* **2016**, *50*, 325–338. [CrossRef]
11. Bossola, M. Nutritional Interventions in Head and Neck Cancer Patients Undergoing Chemoradiotherapy: A Narrative Review. *Nutrients* **2015**, *7*, 266–277. [CrossRef]
12. Sammut, L.; Ward, M.; Patel, N. Physical activity and quality of life in head and neck cancer survivors: A literature review. *Int. J. Sports Med.* **2014**, *35*, 794–799. [CrossRef]
13. Moher, D.; Liberati, A.; Tetzlaff, J.; Altman, D.G.; Group, P. Preferred reporting items for systematic reviews and meta-analyses: The PRISMA statement. *Int. J. Surg.* **2010**, *8*, 336–341. [CrossRef] [PubMed]
14. Macedo, L.G.; Elkins, M.R.; Maher, C.G.; Moseley, A.M.; Herbert, R.D.; Sherrington, C. There was evidence of convergent and construct validity of physiotherapy evidence database quality scale for physiotherapy trials. *J. Clin. Epidemiol.* **2010**, *63*, 920–925. [CrossRef] [PubMed]
15. Yamato, T.P.; Maher, C.; Koes, B.; Moseley, A. The PEDro scale had acceptably high convergent validity, construct validity, and interrater reliability in evaluating methodological quality of pharmaceutical trials. *J. Clin. Epidemiol.* **2017**, *86*, 176–181. [CrossRef]
16. Furuya-Kanamori, L.; Barendregt, J.J.; Doi, S.A. A new improved graphical and quantitative method for detecting bias in meta-analysis. *Int. J. Evid. Based Healthc.* **2018**, *16*, 195–203. [CrossRef]
17. Harris, R.; Bradburn, M.; Deeks, J.; Harbord, R.; Altman, D.; Steichen, T. METAN: Stata module for fixed and random effects meta-analysis. In *Statistical Software Components*; Boston College Department of Economics: Boston, MA, USA, 2010.
18. Sterne, J. METAFUNNEL: Stata module to produce funnel plots for meta-analysis. *Res. Pap. Econ.* **2003**. Available online: https://EconPapers.repec.org/RePEc:boc:bocode:s434101 (accessed on 6 August 2020).

19. Harbord, R.; Harris, R.J.; Sterne, J.A.; Steichen, T. METABIAS: Stata module to test for small-study effects in meta-analysis. *Res. Pap. Econ.* **2009**. Available online: https://EconPapers.repec.org/RePEc:boc:bocode:s404901 (accessed on 10 August 2020).
20. Furuya-Kanamori, L. LFK: Stata module to compute LFK index and Doi plot for detection of publication bias in meta-analysis. *Res. Pap. Econ.* **2020**. Available online: https://EconPapers.repec.org/RePEc:boc:bocode:s458762 (accessed on 14 August 2020).
21. Daly, J.M.; Hearne, B.; Dunaj, J.; LePorte, B.; Vikram, B.; Strong, E.; Green, M.; Muggio, F.; Groshen, S.; DeCosse, J.J. Nutritional rehabilitation in patients with advanced head and neck cancer receiving radiation therapy. *Am. J. Surg.* **1984**, *148*, 514–520. [CrossRef]
22. Samuel, S.R.; Maiya, A.G.; Fernandes, D.J.; Guddattu, V.; Saxena, P.P.; Kurian, J.R.; Lin, P.-J.; Mustian, K.M. Effectiveness of exercise-based rehabilitation on functional capacity and quality of life in head and neck cancer patients receiving chemo-radiotherapy. *Supportive Care Cancer* **2019**, *27*, 3913–3920. [CrossRef]
23. Cereda, E.; Cappello, S.; Colombo, S.; Klersy, C.; Imarisio, I.; Turri, A.; Caraccia, M.; Borioli, V.; Monaco, T.; Benazzo, M.; et al. Nutritional counseling with or without systematic use of oral nutritional supplements in head and neck cancer patients undergoing radiotherapy. *Radiother. Oncol.* **2018**, *126*, 81–88. [CrossRef]
24. Hearne, B.E.; Dunaj, J.M.; Daly, J.M.; Strong, E.W.; Vikram, B.; LePorte, B.J.; DeCosse, J.J. Enteral nutrition support in head and neck cancer: Tube vs. oral feeding during radiation therapy. *J. Am. Diet. Assoc.* **1985**, *85*, 669–674.
25. Isenring, E.; Capra, S.; Bauer, J.; Davies, P.S. The impact of nutrition support on body composition in cancer outpatients receiving radiotherapy. *Acta Diabetol.* **2003**, *40* (Suppl. 1), S162–S164. [CrossRef]
26. Jiang, W.; Ding, H.; Li, W.; Ling, Y.; Hu, C.; Shen, C. Benefits of oral nutritional supplements in patients with locally advanced nasopharyngeal cancer during concurrent chemoradiotherapy: An exploratory prospective randomized trial. *Nutr. Cancer* **2018**, *70*, 1299–1307. [CrossRef]
27. Ravasco, P.; Monteiro-Grillo, I.; Marques Vidal, P.; Camilo, M.E. Impact of nutrition on outcome: A prospective randomized controlled trial in patients with head and neck cancer undergoing radiotherapy. *Head Neck J. Sci. Spec. Head Neck* **2005**, *27*, 659–668. [CrossRef]
28. Rogers, L.Q.; Anton, P.M.; Fogleman, A.; Hopkins-Price, P.; Verhulst, S.; Rao, K.; Malone, J.; Robbs, R.; Courneya, K.S.; Nanavati, P.; et al. Pilot, randomized trial of resistance exercise during radiation therapy for head and neck cancer. *Head Neck* **2013**, *35*, 1178–1188. [CrossRef]
29. Roussel, L.; Micault, E.; Peyronnet, D.; Blanchard, D.; Guarnieri, S.; Choussy, O.; Gery, B.; Bequignon, A.; Joubert, C.; Parienti, J.; et al. Intensive nutritional care for patients treated with radiotherapy in head and neck cancer: A randomized study and meta-analysis. *Eur. Arch. Otorhinolaryngol.* **2017**, *274*, 977–987. [CrossRef] [PubMed]
30. Samuel, S.R.; Maiva, G.A.; Babu, A.S.; Vidyasagar, M.S. Effect of exercise training on functional capacity & quality of life in head & neck cancer patients receiving chemoradiotherapy. *Indian J. Med. Res.* **2013**, *137*, 515–520. [PubMed]
31. Zhao, S.G.; Alexander, N.B.; Djuric, Z.; Zhou, J.; Tao, Y.; Schipper, M.; Feng, F.Y.; Eisbruch, A.; Worden, F.P.; Strath, S.J.; et al. Maintaining physical activity during head and neck cancer treatment: Results of a pilot controlled trial. *Head Neck* **2016**, *38* (Suppl. 1), E1086–E1096. [CrossRef]
32. Capozzi, L.C.; McNeely, M.L.; Lau, H.Y.; Reimer, R.A.; Giese-Davis, J.; Fung, T.S.; Culos-Reed, S.N. Patient-reported outcomes, body composition, and nutrition status in patients with head and neck cancer: Results from an exploratory randomized controlled exercise trial. *Cancer* **2016**, *122*, 1185–1200. [CrossRef]
33. Weaver, C.M.; Miller, J.W. Challenges in conducting clinical nutrition research. *Nutr. Rev.* **2017**, *75*, 491–499. [CrossRef]
34. de Vries, J.; Antoine, J.-M.; Burzykowski, T.; Chiodini, A.; Gibney, M.; Kuhnle, G.; Méheust, A.; Pijls, L.; Rowland, I. Markers for nutrition studies: Review of criteria for the evaluation of markers. *Eur. J. Nutr.* **2013**, *52*, 1685–1699. [CrossRef]
35. Deutz, N.E.; Ashurst, I.; Ballesteros, M.D.; Bear, D.E.; Cruz-Jentoft, A.J.; Genton, L.; Landi, F.; Laviano, A.; Norman, K.; Prado, C.M. The underappreciated role of low muscle mass in the management of malnutrition. *J. Am. Med. Dir. Assoc.* **2019**, *20*, 22–27. [CrossRef]
36. Prado, C.M.; Purcell, S.A.; Alish, C.; Pereira, S.L.; Deutz, N.E.; Heyland, D.K.; Goodpaster, B.H.; Tappenden, K.A.; Heymsfield, S.B. Implications of low muscle mass across the continuum of care: A narrative review. *Ann. Med.* **2018**, *50*, 675–693. [CrossRef] [PubMed]

37. Nilsen, T.S.; Scott, J.M.; Michalski, M.; Capaci, C.; Thomas, S.; Herndon, J.E. Novel methods for reporting of exercise dose and adherence: An exploratory analysis. *Med. Sci. Sports Exerc.* **2018**, *50*, 1134. [CrossRef]
38. Fayh, A.P.T.; de Sousa, I.M.; Gonzalez, M.C. New insights on how and where to measure muscle mass. *Curr. Opin. Supportive Palliat. Care* **2020**. [CrossRef]
39. Vasiloglou, M.F.; Fletcher, J.; Poulia, K.-A. Challenges and perspectives in nutritional counselling and nursing: A narrative review. *J. Clin. Med.* **2019**, *8*, 1489. [CrossRef]
40. McCarter, K.; Baker, A.L.; Britton, B.; Halpin, S.A.; Beck, A.; Carter, G.; Wratten, C.; Bauer, J.; Wolfenden, L.; Burchell, K. Head and neck cancer patient experience of a new dietitian-delivered health behaviour intervention: 'you know you have to eat to survive'. *Supportive Care Cancer* **2018**, *26*, 2167–2175. [CrossRef]
41. Endevelt, R.; Gesser-Edelsburg, A. A qualitative study of adherence to nutritional treatment: Perspectives of patients and dietitians. *Patient Prefer. Adherence* **2014**, *8*, 147. [CrossRef]
42. Ferreira, I.B.; Lima, E.d.N.S.; Canto, P.P.L.; Gontijo, C.A.; Maia, Y.C.d.P.; Pena, G.d.G. Oral nutritional supplementation affects the dietary intake and body weight of head and neck cancer patients during (Chemo) radiotherapy. *Nutrients* **2020**, *12*, 2516. [CrossRef]
43. Sladdin, I.; Ball, L.; Bull, C.; Chaboyer, W. Patient-centred care to improve dietetic practice: An integrative review. *J. Hum. Nutr. Diet.* **2017**, *30*, 453–470. [CrossRef]
44. Williams, G.F.; White, H.; Sen, M.; Prestwich, R.J.D. Patients' experience of enteral feeding following (chemo) radiotherapy for head and neck cancer: A qualitative study. *Clin. Nutr.* **2019**, *38*, 1382–1389. [CrossRef]
45. Sandmæl, J.A.; Sand, K.; Bye, A.; Solheim, T.S.; Oldervoll, L.; Helvik, A.S. Nutritional experiences in head and neck cancer patients. *Eur. J. Cancer Care* **2019**, *28*, e13168. [CrossRef]
46. Orell, H.; Schwab, U.; Saarilahti, K.; Österlund, P.; Ravasco, P.; Mäkitie, A. Nutritional counseling for head and neck cancer patients undergoing (chemo) radiotherapy—A prospective randomized trial. *Front. Nutr.* **2019**, *6*, 22. [CrossRef] [PubMed]
47. Campbell, K.L.; Winters-Stone, K.M.; Wiskemann, J.; May, A.M.; Schwartz, A.L.; Courneya, K.S.; Zucker, D.S.; Matthews, C.E.; Ligibel, J.A.; Gerber, L.H. Exercise guidelines for cancer survivors: Consensus statement from international multidisciplinary roundtable. *Med. Sci. Sports Exerc.* **2019**, *51*, 2375–2390. [CrossRef] [PubMed]
48. Sweegers, M.G.; Altenburg, T.M.; Chinapaw, M.J.; Kalter, J.; Verdonck-de Leeuw, I.M.; Courneya, K.S.; Newton, R.U.; Aaronson, N.K.; Jacobsen, P.B.; Brug, J. Which exercise prescriptions improve quality of life and physical function in patients with cancer during and following treatment? A systematic review and meta-analysis of randomised controlled trials. *Br. J. Sports Med.* **2018**, *52*, 505–513. [CrossRef] [PubMed]

nutrients

MDPI

Article

Computed-Tomography Body Composition Analysis Complements Pre-Operative Nutrition Screening in Colorectal Cancer Patients on an Enhanced Recovery after Surgery Pathway

Pamela Klassen [1], Vickie Baracos [2], Leah Gramlich [3], Gregg Nelson [4], Vera Mazurak [1,*] and Lisa Martin [2]

[1] Department of Agricultural, Food & Nutritional Sciences, University of Alberta, Edmonton, AB T6G2P5, Canada; pamela2@ualberta.ca
[2] Department of Oncology, University of Alberta, Edmonton, AB T6G2P5, Canada; vickie.baracos@ualberta.ca (V.B.); ls2@ualberta.ca (L.M.)
[3] Department of Medicine, University of Alberta, Edmonton, AB T6G2P5, Canada; lg3@ualberta.ca
[4] Department of Oncology, University of Calgary, Calgary, AB T6G2P5, Canada; Gregg.Nelson@albertahealthservices.ca
* Correspondence: vera.mazurak@ualberta.ca

Received: 30 October 2020; Accepted: 3 December 2020; Published: 5 December 2020

Abstract: Pre-operative nutrition screening is recommended to identify cancer patients at risk of malnutrition, which is associated with poor outcomes. Low muscle mass (sarcopenia) and lipid infiltration to muscle cells (myosteatosis) are similarly associated with poor outcomes but are not routinely screened for. We investigated the prevalence of sarcopenia and myosteatosis across the nutrition screening triage categories of the Patient-Generated Subjective Global Assessment Short Form (PG-SGA$_{SF}$) in a pre-operative colorectal cancer (CRC) cohort. Data were prospectively collected from patients scheduled for surgery at two sites in Edmonton, Canada. PG-SGA$_{SF}$ scores $\geq$ 4 identified patients at risk for malnutrition; sarcopenia and myosteatosis were identified using computed-tomography (CT) analysis. Patients (n = 176) with a mean age of 63.8 ± 12.0 years, 52.3% male, 90.3% with stage I–III disease were included. Overall, 25.2% had PG-SGA$_{SF}$ score $\geq$ 4. Sarcopenia alone, myosteatosis alone or both were identified in 14.0%, 27.3%, and 6.4% of patients, respectively. Sarcopenia and/or myosteatosis were identified in 43.4% of those with PG-SGA$_{SF}$ score < 4 and in 58.5% of those with score $\geq$ 4. Overall, 32.9% of the cohort had sarcopenia and/or myosteatosis with PG-SGA$_{SF}$ score < 4. CT-defined sarcopenia and myosteatosis are prevalent in pre-operative CRC patients, regardless of the presence of traditional nutrition risk factors (weight loss, problems eating); therefore, CT image analysis effectively adds value to nutrition screening by identifying patients with other risk factors for poor outcomes.

Keywords: malnutrition; sarcopenia; colorectal cancer; CT; PG-SGA; subjective global assessment; myosteatosis; muscle mass

1. Introduction

Globally, colorectal cancer is among the most frequently diagnosed cancers accounting for 1.8 million new diagnoses and 800,000 deaths in 2018 [1]. Surgical resection typically occurs shortly after diagnosis especially in organ-confined disease. Depending on risk factors, resection may be followed by adjuvant chemotherapy to reduce recurrence risk. In locally advanced disease, surgical resection may be preceded by neo-adjuvant chemotherapy and/or radiation and subsequently followed by adjuvant chemotherapy. In these cases, total curative treatment time can be up to 18 months.

In both surgical and oncological contexts, routine nutrition screening is recommended to identify patients with or at risk of malnutrition, characterized in cancer patients by negative energy balance and skeletal muscle loss [2–4]. Malnutrition is associated with longer length of post-operative hospital stay [5], surgical complications [6–9], and reduced overall survival for cancer patients [10–12], therefore early identification of risk factors that can lead to malnutrition (e.g., weight loss, problems eating, poor appetite) is essential. Pre-operative nutritional care, starting with nutrition screening, is one of the tenets of the evidence-based, multi-modal Enhanced Recovery After Surgery (ERAS) protocol. Designed to reduce peri-operative stress, maintain physiological function post-operatively, and promote faster recovery, the widely-accepted protocol consists of approximately 20 recommendations including several related to optimal nutrition care from pre-operative nutrition screening to post-operative early feeding and immuno-nutrition [3]. However, despite general agreement on the benefit of nutrition screening in both surgical and oncological settings, no single screening tool is recommended [2–4]. The Patient-Generated Subjective Global Assessment Short Form (PG-SGA$_{SF}$, © FD Ottery, 2001) is a validated screening tool that is commonly used in ambulatory oncology settings. As an abridged version of the patient and clinician-completed Patient-Generated Subjective Global Assessment, it uses patient-reported recent weight loss, symptoms, and difficulty eating as indicators of risk to quickly identify patients who may benefit from further nutritional assessment and intervention [13,14].

While traditional indicators of nutrition risk such as weight loss, low BMI, and low oral intake certainly increase risk of malnutrition and associated poor outcomes, evidence has accumulated that features such as computed tomography (CT)-defined low skeletal muscle mass and fat infiltration to the muscle cells are also strongly associated with negative clinical outcomes such as increased risk of post-operative complications [6,15,16], longer post-operative length of stay [17] and reduced overall survival [10,16,18–21] in oncology patients. In oncology patients, low skeletal muscle mass associated with poor clinical outcomes is referred to as sarcopenia [22]. Although there are various methods of assessing skeletal muscle, CT analysis of cross-sectional images at the third lumbar (L3) vertebrae is considered the gold standard in the oncological setting to precisely quantify skeletal muscle. Not only are CT images routinely available for the majority of patients, this assessment carries no additional patient burden and can be completed prior to an in-person assessment [23]. Other methods of identifying muscle loss do exist, including the full PG-SGA (clinician-completed), which includes a nutrition-focused physical exam [24]. However, sarcopenia can be masked by overweight and obesity and therefore go undetected on physical exams or nutrition risk screening [25,26]. Similarly, sarcopenia-specific screening tools such as SARC-F can be used to identify patients who have functional changes as a result of low muscle mass [27], but does not quantify the muscle. Both tools require additional patient time, contact and clinic space, and neither tool allows for accurate visualization of the muscle.

With the prolific use of CT body composition analysis in oncology research, low muscle radiodensity (known as myosteatosis) has emerged alongside sarcopenia as an additional prognostic factor. Skeletal muscle radiodensity (SMR, reported in Hounsfield Units, HU) inversely reflects the triglyceride content of skeletal muscle. Low SMR is often referred to as myosteatosis; thresholds for which have been defined according to associations with overall survival after chemotherapy [11] as well as length of stay and hospital readmissions after colorectal cancer surgery [17]. Myosteatosis, as a characteristic of muscle, cannot be identified by any method other than CT analysis.

Since nutrition screening is already part of the pre-operative ERAS pathway, and in light of the known impacts of malnutrition, sarcopenia and myosteatosis on surgical and oncological outcomes, it is relevant to explore the prevalence of sarcopenia and myosteatosis across different levels of nutrition screening results. For advanced cancer patients, prior work confirms that sarcopenia and myosteatosis are prevalent across all levels of nutrition risk (low to high) using a variety of nutrition screening tools [11,28]. In early stage cancer patients, the prevalence of sarcopenia and myosteatosis has been described but not been analyzed in tandem with nutrition screening results [17]. Identifying whether high nutrition screening scores consistently co-exist with sarcopenia and myosteatosis in early stage disease will inform the development of care pathways to ensure that all relevant risk factors are identified early. In the present study, we aimed to describe the prevalence of sarcopenia and myosteatosis according to level of nutrition risk as defined by the PG-SGA$_{SF}$ in patients with CRC presenting for elective surgical resection. We hypothesized that there would be a high prevalence of sarcopenia and myosteatosis across all of the nutrition risk triage categories of the PG-SGA$_{SF}$.

2. Materials and Methods

Data were prospectively collected from consecutive patients ≥18 years old presenting for pre-operative assessment prior to elective surgical resection of a primary CRC. Data collection occurred at two acute care centres in Edmonton, Alberta between 2016 and 2017; both centres had implemented the Enhanced Recovery After Surgery (ERAS) protocol for colorectal surgery. Patients completed a PG-SGA$_{SF}$ at their first pre-operative visit and the forms were scored by a trained researcher. Patient demographics, cancer stage and cancer site were obtained from the Alberta Cancer Registry. Surgical data (e.g., surgery date and procedure) were obtained from the ERAS Interactive Audit System (EIAS), which has been previously described [5]. Finally, CT images were obtained from the regional Picture Archiving Communication System. Patients were eligible for inclusion if they had completed a pre-operative PG-SGA$_{SF}$, had a confirmed diagnosis and cancer stage, had an analyzable CT image within 6 months prior to surgery, and had complete surgical data available in EIAS. Ethical approval was granted from the local health research ethics board (protocol identifier: HREBA.CC-16-0308).

The PG-SGA$_{SF}$ consists of scored patient-reported components including weight change, changes in food intake, symptoms impacting the ability to eat, and performance status; these scores are summed for a total possible score out of 37 (a higher score represents greater nutrition risk). A nutrition triage recommendation is assigned based on the total PG-SGA$_{SF}$ score as follows: no intervention required (scores 0–1), education or pharmaceutical intervention (scores 2–3), registered dietitian intervention (scores 4–8), or critical nutrient intervention and improved symptom management (scores ≥ 9). For the purposes of this analysis, patients were divided by triage recommendation into two groups—scores 0–3 and scores ≥ 4, with the latter group considered to be at risk for malnutrition.

Body composition was analyzed using CT image analysis, previously validated for use in the cancer population [23]. CT body composition analysis makes opportunistic use of existing CT images, taken in this case as part of the normal staging process for patients with colorectal tumors. Cross-sectional skeletal muscle and adipose tissue areas from a single axial image at the third lumbar vertebrae (L3) are highly correlated with total body skeletal muscle and adipose tissue [29]. L3 images were analyzed using an auto-segmentation module (ABACS module, Voroni Health Analytics; Slice-O-Matic©, Tomovision, Montreal, QC, Canada) and manually corrected by a trained technician. Muscle and adipose tissue cross-sectional areas were delineated in cm^2 using pre-defined thresholds; −29 to +150 HU for skeletal muscle, −150 to −50 HU for visceral adipose tissue, and −190 to −30 HU for subcutaneous adipose tissue. The resultant cross-sectional areas were normalized for height and reported as skeletal muscle index (SMI), visceral adipose tissue index (VATI), and subcutaneous adipose tissue index (SATI) in cm^2/m^2. The mean SMR for the entire muscle area at L3 was recorded.

Sarcopenia and myosteatosis were defined using previously identified sex- and age-specific SMI and SMR thresholds, which were developed by Martin et al. based on associations with post-operative length of stay in a large cohort of CRC patients undergoing surgery [17]. The cohort of the present study was similar to the cohort of Martin et al., and therefore these thresholds were deemed highly appropriate. The primary outcome of the present study was to determine the prevalence of sarcopenia and myosteatosis across nutrition risk screening categories in pre-operative colorectal cancer patients and therefore was intended as a descriptive study; it was not powered to evaluate the association between these muscle features and surgical outcomes.

Statistical analysis was performed using IBM SPSS Statistics for Windows, version 24 (SPSS, Chicago, IL, USA). Frequency and summary data are presented, with comparisons between groups analyzed using chi square tests with Bonferroni corrections or independent t-tests where appropriate. A p-value of <0.05 was considered statistically significant.

3. Results

3.1. Patient Characteristics

Completed PG-SGA$_{SF}$, analyzable CT and EIAS data were obtained for 176 patients (Table 1). The sample contained similar proportions of males and females, with a mean age of 63.8 ± 12.0 years. Similar proportions of colon and rectal tumor sites were included, and 90% of patients had stage I–III disease. While data collection aimed to include patients presenting for curative intent surgery, a small proportion were subsequently found to have stage IV disease. Mean BMI was 28.4 ± 6.3 kg/m^2, with no significant difference between males and females. Overweight (BMI of 25–29.9 kg/m^2) and obesity (BMI ≥ 30 kg/m^2) were prevalent, with 42% and 30% of the cohort in these categories, respectively.

Table 1. Characteristics of pre-operative colorectal cancer patients.

Demographics	Male	Female	All (*N* = 176)
Age (years), mean (±SD)	63.6 (10.7)	63.9 (13.3)	63.8 (12.0)
Sex, *N* (%)	92 (52.3)	84 (47.7)	
Tumor site, N (%)			
colon	45 (48.9)	47 (45.0)	92 (52.3)
rectum	47 (51.1)	37 (44.0)	84 (47.7)
Cancer stage, N (%)			
Stage I–II	51 (55.4)	46 (54.7)	97 (55.1)
Stage III	32 (34.8)	30 (35.7)	62 (35.2)
Stage IV	5 (5.4)	5 (6.0)	10 (5.7)
Anthropometrics			
Weight, kg, mean (±SD)	90.7 (18.2)	69.9 (17.2)	80.7 (20.5)
Height, cm, mean (±SD)	176.5 (7.0)	159.0 (8.6)	168.3 (11.7)
BMI, kg/m^2, mean (±SD)	29.0 (5.0)	27.8 (7.5)	28.4 (6.3)
BMI category, kg/m^2, N (%)			
<20	2 (2.2)	9 (11.1)	11 (6.5)
20–24.9	15 (16.9)	21 (25.9)	36 (21.2)
25–29.9	41 (46.1)	31 (38.3)	72 (42.4)
30–34.9	19 (21.3)	10 (12.3)	29 (17.1)
35–39.9	10 (11.2)	4 (4.9)	14 (8.2)
≥40	2 (2.2)	6 (7.4)	8 (4.7)
Body composition by CT analysis			
Mean skeletal muscle index (SMI), cm^2/m^2	53.3 (9.8)	40.9 (7.7)	47.4 (10.8)
Mean skeletal muscle radiodensity (SMR), HU	35.5 (9.1)	35.7 (9.3)	35.6 (9.2)
Subcutaneous adipose tissue index (SATI, cm^2/m^2), mean	67.1 (29.0)	103.6 (58.0)	84.1 (48.4)
Visceral adipose tissue index (VATI, cm^2/m^2), mean	79.7 (38.1)	46.9 (36.9)	64.3 (40.9)
Sarcopenia, myosteatosis or both, *N* (%)	45 (49.5)	38 (46.3)	83 (48.0)
Sarcopenia alone, N (%)	*8 (8.8)*	*16 (19.8) **	*24 (14.0)*
Myosteatosis alone, N (%)	*28 (30.8)*	*19 (23.5)*	*47 (27.3)*
Sarcopenia and Myosteatosis, *N* (%)	9 (9.9) *	2 (2.5)	11 (6.4)
No sarcopenia or myosteatosis, *N* (%)	46 (50.5)	44 (53.7)	90 (52.0)

Cancer Stage: American Joint Committee on Cancer 7th Edition; * $p < 0.05$.

3.2. Nutrition Risk Factors by Patient-Generated Subjective Global Assessment Short Form

Results from the PG-SGA$_{SF}$ are presented in Table 2. On average, patients presented with minimal weight change over the past month (mean −0.4% ± 3.4%); however, there was wide variability, ranging from 30% weight loss to 13% weight gain. The PG-SGA$_{SF}$ scores the severity of weight loss according to five categories, and our patients with 1 month weight change reported were categorized as follows: 0–1.9% weight loss, 129 (83.2%); 2–2.9% weight loss, 10 (6.5%); 3–4.9% weight loss, 9 (5.8%); 5–9.9% weight loss, 5 (3.2%); and ≥10% weight loss, 2 (1.3%).

Table 2. Nutritional risk factors by PG-SGA$_{SF}$.

Domain	Overall, *N* = 176
Box 1: Weight Change	
Weight change past month, mean % (±SD)	−0.4 (3.4)
Weight change past 6 months, mean % (±SD)	−2.0 (5.5)
No change/increased weight in past 2 weeks, *N* (%)	123 (69.9)
Decreased weight in past 2 weeks, *N* (%)	53 (30.1)
Box 2: Food Intake	
Food intake past month, N (%)	
Unchanged/more than usual	149 (84.7)
Less than usual	27 (15.3)
Type of food intake, N (%)	
Normal food, normal amount	143 (81.3)
Normal food, less than normal amount	18 (10.2)
Little solid food	3 (1.7)
Only liquids or nutritional supplements	8 (4.5)
Very little of anything	4 (2.3)
Only tube feeding/feeding by vein	0 (0)
Box 3: Nutrition Impact Symptoms, *N* (%)	
No problems eating	150 (85.2)
No appetite	14 (8.0)
Nausea	7 (4.0)
Constipation	14 (8.0)
Diarrhea	18 (10.2)
Vomiting	3 (1.7)
Feel full quickly	6 (3.4)
Foods taste funny or have no taste	3 (1.7)
Smells bother me	3 (1.7)
Mouth sores	0 (0)
Problem swallowing	3 (1.7)
Fatigue	16 (9.1)
Pain	4 (2.3)
Dry mouth	7 (4.0)
Other	4 (2.3)
Box 4: Activity and Function, *N* (%)	
Normal, no limitations	115 (65.3)
Not normal self, fairly normal activities	44 (25.0)
Not feeling up to most things, in bed or chair <half day	8 (4.5)
Not able to do most things or pretty much bedridden	7 (4.0)

PG-SGA$_{SF}$, Patient-Generated Subjective Global Assessment Short Form.

Food intake was unchanged for 84.7% of patients, and 85.2% reported no problems eating. The most frequent nutrition impact symptoms included diarrhea (10.2%), fatigue (9.1%), no appetite (8.0%) and constipation (8.0%). Finally, the vast majority of patients reported normal or fairly normal activity.

Overall, the mean total PG-SGA$_{SF}$ score was 2.9 ± 4.8 (Table 3), out of a total possible score of 37; scores ranged from 0 to 26. Patients with scores of 0-1 comprised 59.8% of the cohort, indicating no intervention required, and 25.2% of patients scored ≥ 4, indicating a need for dietitian assessment or intervention. Among patients who scored ≥ 4, mean total score was 9.2 ± 5.8.

Table 3. Patient-Generated Subjective Global Assessment Short Form scores and triage recommendation.

PG-SGA$_{SF}$ Domain	Score 0–3	Score ≥ 4	Overall
Box 1: Weight Change (max. 5; mean ± SD)	0.27 (0.64)	1.77 (1.38)	0.64 (1.10)
Box 2: Food Intake (max. 5; mean ± SD)	0.13 (0.53)	1.68 (1.74)	0.52 (1.19)
Box 3: Nutrition Impact Symptoms (max. 24; mean ± SD)	0.08 (0.41)	4.59 (4.30)	1.20 (2.92)
Box 4: Activity and Function (max. 3; mean ± SD)	0.22 (0.50)	1.18 (0.95)	0.47 (0.77)
Total Score, mean ± SD	0.70 (1.05)	9.23 (5.77)	2.86 (4.79)
Triage Recommendation			*N*, %
0–1 (no intervention, reassess regularly)			104 (59.8)
2–3 (patient/family education; pharmacological intervention as indicated by symptoms)			26 (14.9)
4–8 (intervention by RD and nurse or physician as indicated by symptoms)			26 (14.9)
≥9 (critical need for symptom management and nutrition intervention)			18 (10.3)

RD, registered dietitian; max., maximum score possible.

3.3. CT-Defined Skeletal Muscle Analysis

Mean SMI for males and females were 53.3 ± 9.8 and 40.9 ± 7.7 cm^2/m^2, respectively, with a mean of 47.4 ± 10.8 cm^2/m^2 overall. SMR for males and females was 35.5 ± 9.1 and 35.7 ± 9.3 HU, respectively, with an overall mean of 35.6 ± 9.2 HU. The distributions of these features were consistent with the analysis of a large cohort (N = 2100), similar CRC cohort described by Martin et al., shown in Figure 1a and b. Based on the thresholds developed by Martin et al. [17] (Figure 1c), sarcopenia alone was identified in 14.0%, myosteatosis alone in 27.3%, and the combination of both in an additional 6.4% (Table 1) in our cohort. More females than males presented with sarcopenia alone (19.8% vs. 8.8%, $p < 0.05$), and more males than females presented with combined sarcopenia and myosteatosis (9.9% vs. 2.5%; $p < 0.05$). However, there was no difference between sexes in the overall prevalence of sarcopenia and/or myosteatosis (49.5% of males vs. 46.3% of females).

3.4. Co-Existence of Nutrition Risk by PG-SGA$_{SF}$ and CT-Defined Sarcopenia and Myosteatosis

The prevalence of sarcopenia and myosteatosis according to PG-SGA$_{SF}$ triage category was evaluated. Sarcopenia and/or myosteatosis were prevalent both above and below a PG-SGA$_{SF}$ cutoff score of ≥4. Of the patients with scores < 4 (75.9% of the cohort), 43.4% had CT-defined sarcopenia and/or myosteatosis, compared to 58.5% of those with screening scores ≥ 4, with no significant difference between groups, as shown in Table 4. Of the entire cohort, only 14.1% had co-existing PG-SGA$_{SF}$ score ≥ 4 and sarcopenia or myosteatosis. One-third (32.9%) had sarcopenia or myosteatosis with PG-SGA$_{SF}$ score < 4 (Figure 2), and therefore were identified by nutrition screening as having low risk of malnutrition but were found to have sarcopenia or myosteatosis using CT analysis.

Table 4. Prevalence of sarcopenia and myosteatosis by PG-SGA$_{SF}$ score.

Characteristic	PG-SGA$_{SF}$ Score 0–3	PG-SGA$_{SF}$ Score ≥ 4
Sarcopenia alone, *N* (%)	15 (11.6)	7 (17.1)
Myosteatosis alone, *N* (%)	30 (23.3)	17 (41.5)
Sarcopenia and myosteatosis, *N* (%)	11 (8.5)	0

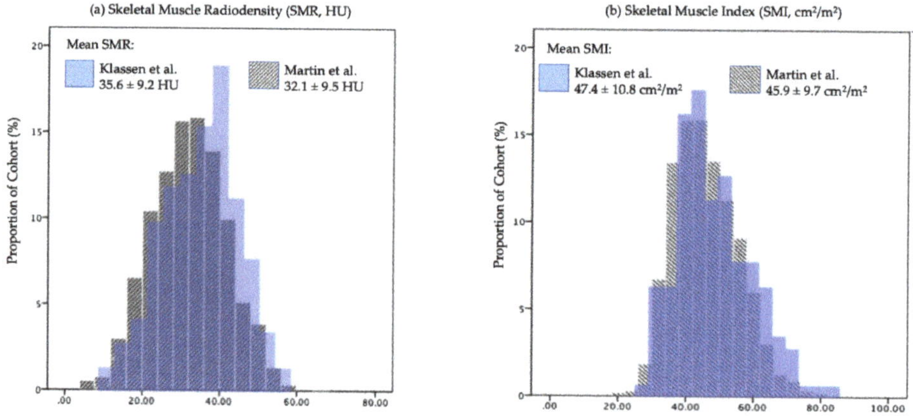

(c) Sex- and age-specific thresholds used to define sarcopenia and myosteatosis by Martin et al. [17]

		<50 years	50-59 years	60-69 years	70-79 years	≥80 years
Sarcopenia	Male	<50.6	<49.3	<46.8	<43.4	<38.7
(SMI, cm²/m²)	Female	<39.6	<37.6	<37.1	<35.2	<33.5
Myosteatosis	Male	<42.0	<37.6	<32.9	<29.7	<28.1
(SMR, HU)	Female	<39.6	<36.0	<31.4	<27.8	<24.8

Figure 1. (**a**) and (**b**). Similar distributions of (**a**) skeletal muscle radiodensity and (**b**) skeletal muscle index from two pre-operative colorectal cancer cohorts: the present cohort ($N = 176$) and Martin et al. ($N = 2100$) [17]. (**c**). Martin et al. thresholds for sarcopenia and myosteatosis using skeletal muscle index (SMI) and skeletal muscle radiodensity (SMR), respectively, developed based on associations with longer post-operative length of stay in early-stage colorectal cancer patients [17].

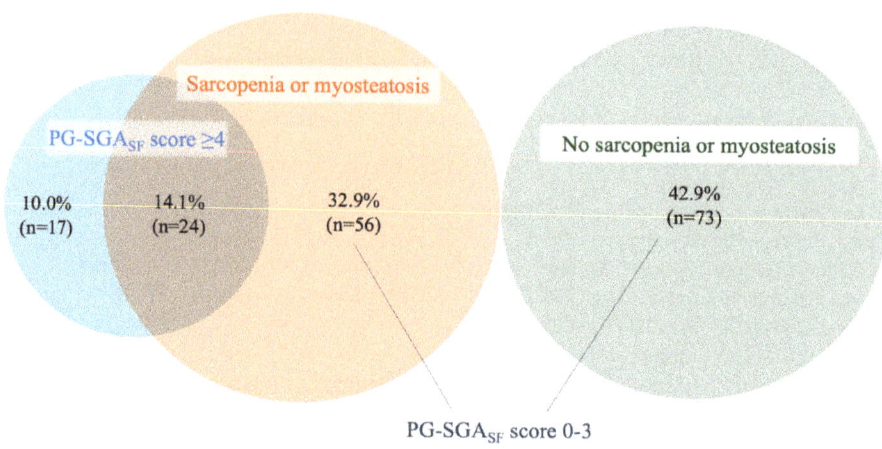

Figure 2. Proportional Venn diagram illustrating the co-existence of PG-SGA$_{SF}$ score ≥ 4 and sarcopenia or myosteatosis pre-operatively. Sarcopenia or myosteatosis co-existed with a PG-SGA$_{SF}$ score of 0–3 in 32.9% of the total cohort.

4. Discussion

4.1. Early Nutrition Risk Exists

In this pre-operative, CRC cohort treated with curative intent, one-quarter of patients presented with PG-SGA$_{SF}$ scores ≥ 4 early in their treatment journey. This cohort represents a demographic that is not well-characterized in the literature with respect to nutrition screening than patients with unresectable or metastatic disease, for whom the prevalence of nutrition risk ranges from 36 to 64% (i.e., PG-SGA$_{SF}$ score ≥ 4 or SGA B/C) [6,11,13,30]. The lower prevalence of PG-SGA$_{SF}$ score ≥ 4 in our cohort compared to cohorts with advanced disease may reflect a smaller window of time for weight loss and functional decline to occur from disease onset to assessment, and that nutrition impact symptoms are less common prior to chemotherapy treatment. However, it is likely that early malnutrition will progress over the course of treatment due to the stress of surgery and nutrition impact symptoms caused by adjuvant chemotherapy [31]. For these patients, the pre- and peri-operative periods are essential times to intervene and attempt to halt the progression of malnutrition [32].

4.2. *Sarcopenia and Myosteatosis Are Prevalent in Pre-Operative CRC Patients*

Sarcopenia and myosteatosis were prevalent across PG-SGA$_{SF}$ triage categories and regardless of sex, with 48% of the cohort having at least one of these muscle features. While sarcopenia and myosteatosis do not necessarily occur together, both are independently associated with poor outcomes [17]. In the short term, pre-operative sarcopenia and myosteatosis carry increased risks of post-operative complications [6,15], longer length of stay and hospital readmission [17]. Patients with sarcopenia at diagnosis who continue to lose muscle over the next two years have significantly worse overall survival than those with stable or increased muscle mass [33]. Our prevalence data suggest that sarcopenia and myosteatosis are less common early in the disease trajectory compared to advanced stage but are still widespread. For example, Ní Bhuachalla et al. analyzed 725 CRC patients, 45% with stage IV disease, and found 41% to have sarcopenia and 46% to have myosteatosis, compared to our results of 20 % and 34%, respectively [28]. Yet, most patients with stage II–III CRC require adjuvant chemotherapy after surgery or future chemotherapy upon recurrence, where sarcopenia and myosteatosis are known to be associated with increased treatment toxicity, poor quality of life and decline in functional status [18,19]. Although nutrition elements exist within the ERAS pathway, nutrition screening alone without body composition analysis remains the primary method of identifying patients requiring nutrition intervention [3]. Early identification of sarcopenia and myosteatosis could enable prompt intervention, prevent progressive muscle degradation, and further optimize outcomes.

4.3. *Nutrition Risk and Skeletal Muscle Aberrations Are Distinct Risk Factors*

This co-analysis of CT-defined sarcopenia and myosteatosis alongside PG-SGA$_{SF}$ nutrition screening corroborates the work of others showing that low nutrition screening scores do not preclude risk for poor outcomes conferred by sarcopenia and myosteatosis [11,12,28]. Despite three-quarters of our cohort having low PG-SGA$_{SF}$ screening scores (0–3), nearly half of these had pre-existing sarcopenia and/or, myosteatosis. CT analysis, therefore, identified one-third (32.9%) of the total cohort as having sarcopenia and/or myosteatosis early in the treatment journey, in the absence of traditional nutrition risk factors. In light of this, we suggest that patients with muscle aberration are a distinct population, requiring unique screening tools. In early stage disease, sarcopenia and myosteatosis may not be related to reduced oral intake, which the PG-SGA$_{SF}$ would identify, but rather may be pre-existing and related to low baseline activity levels, poor diet quality, or comorbid conditions such as COPD or diabetes, as described by Xiao et al [34]. This recognition does not discount the positive predictive value of nutrition screening, but rather emphasizes the need for complementary screening in early stage cancer to identify patients with sarcopenia and myosteatosis. Furthermore, efforts are required to identify the risk factors predisposing patients to these aberrations, and particularly those which are amendable to intervention.

4.4. *Enhancing Identification of At-Risk Patients*

Our results confirm that pre-operative CRC patients commonly have sarcopenia or myosteatosis in the absence of reductions in food intake and weight, and therefore may not be selected for early intervention using traditional nutrition screening tools. The pre-operative consultation is an opportune time to assess patients not only for weight loss or change in oral intake, but also for existing sarcopenia and myosteatosis. CT analysis is the most suitable tool in this population, given that 73% of our cohort was overweight or obese, with a significant number having myosteatosis which is otherwise not identified by physical or functional assessments. Furthermore, CT analysis avoids increasing the patient time requirement in the screening process and requires similar or less time from a clinician than an in-person assessment. The limitations of CT analysis should be acknowledged, including the requirement for trained personnel, the cost of for-purpose software, and the requirement for CT images at the L3 vertebrae to be available. These limitations, however, are quickly being addressed. In the oncology context, CT images are routinely available. Furthermore, free versions of the analysis software are now widely used, and efforts are ongoing to validate analysis at other locations such as first lumbar vertebrae and twelfth thoracic vertebrae in populations without abdominal CT scans [35].

Future work in this area should focus on the feasibility of integrating screening for sarcopenia and myosteatosis with nutrition screening processes, thereby accurately identifying both risk factors. A clinical flow for surgical oncology that includes PG-SGA$_{SF}$ screening followed by CT analysis of body composition for patients with low screening scores (Figure 3) would effectively capture patients with traditional malnutrition risk factors and those with sarcopenia or myosteatosis. Further detailed assessment after this screening is warranted, and could include dietary assessment or functional assessments depending on the setting. In non-oncology settings, the use of an alternative screening tool such as SARC-F to complement nutrition screening may similarly be useful to identify sarcopenia [36], despite its inability to detect myosteatosis. In either case, the improved identification of these distinct groups of at-risk patients will set the stage for research into effective interventions, improved clinical care pathways and multi-modal pre-habilitation programs using exercise, nutrition and psychosocial support [37–41] which have yet to be formally included in ERAS programs. Through a combination of enhanced screening and targeted interventions, the pre-operative period can become an opportunity to enhance outcomes from cancer surgery through survivorship.

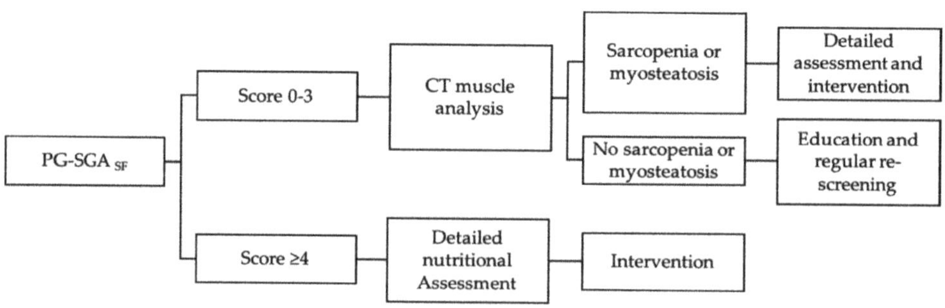

Figure 3. Proposed clinical flow for surgical oncology patients starting with nutrition screening, followed by CT muscle analysis for patients with low screening scores, to identify patients with skeletal muscle aberration in the absence of overt nutrition risk.

Author Contributions: V.M., G.N., L.G. and L.M. conceived the study design; V.B. and L.M. curated the data; P.K. performed the analysis; V.B. resources, V.M. and L.M. provided supervision, V.M., L.M. and P.K. provided the visualization, P.K. drafted the manuscript, V.M., V.B., L.M., P.K., L.G. and G.N. reviewed and edited the manuscript. All authors have read and agreed to the published version of the manuscript.

Funding: Pamela Klassen is the recipient of the 2019 Alberta SPOR Graduate Studentship in Patient-Oriented Research. The Alberta SPOR SUPPORT Unit is jointly funded by Alberta Innovates and the Canadian Institute of Health Research. Lisa Martin receives support from a CIHR-CNS fellowship.

Acknowledgments: The authors would like to acknowledge Abha Dunichand-Hoedl for her contribution to CT image analysis.

Conflicts of Interest: The authors declare no conflict of interest.

References

1. Bray, F.; Ferlay, J.; Soerjomataram, I.; Siegel, R.L.; Torre, L.A.; Jemal, A. Global cancer statistics 2018: GLOBOCAN estimates of incidence and mortality worldwide for 36 cancers in 185 countries. *CA Cancer J. Clin.* **2018**, *68*, 394–424. [CrossRef] [PubMed]
2. Arends, J.; Bachmann, P.; Baracos, V.; Barthelemy, N.; Bertz, H.; Bozzetti, F.; Fearon, K.; Hütterer, E.; Isenring, E.; Kaasa, S.; et al. ESPEN guidelines on nutrition in cancer patients. *Clin. Nutr.* **2016**, *36*, 11–48. [CrossRef] [PubMed]
3. Gustafsson, U.O.; Scott, M.J.; Hubner, M.; Nygren, J.; Demartines, N.; Francis, N.; Rockall, T.A.; Young-Fadok, T.M.; Hill, A.G.; Soop, M.; et al. Guidelines for Perioperative Care in Elective Colorectal Surgery: Enhanced Recovery After Surgery (ERAS ®) Society Recommendations: 2018. *World J. Surg.* **2019**, *43*, 659–695. [CrossRef] [PubMed]
4. Thompson, K.L.; Elliott, L.; Fuchs-tarlovsky, V.; Levin, R.M.; Voss, A.C.; Piemonte, T.; Ld, N. Guideline for Adults. *J. Acad. Nutr. Diet.* **2017**, *117*, 297–310.e47. [CrossRef]
5. Martin, L.; Gillis, C.; Atkins, M.; Gillam, M.; Sheppard, C.; Buhler, S.; Hammond, C.B.; Nelson, G.; Gramlich, L. Implementation of an Enhanced Recovery After Surgery Program Can Change Nutrition Care Practice: A Multicenter Experience in Elective Colorectal Surgery. *J. Parenter. Enter. Nutr.* **2019**, *43*, 206–219. [CrossRef]
6. Maurício, S.F.; Xiao, J.; Prado, C.M.; Gonzalez, M.C.; Correia, M.I.T.D. Different nutritional assessment tools as predictors of postoperative complications in patients undergoing colorectal cancer resection. *Clin. Nutr.* **2018**, *37*, 1505–1511. [CrossRef]
7. Schwegler, I.; Von Holzen, A.; Gutzwiller, J.P.; Schlumpf, R.; Mühlebach, S.; Stanga, Z. Nutritional risk is a clinical predictor of postoperative mortality and morbidity in surgery for colorectal cancer. *Br. J. Surg.* **2010**, *97*, 92–97. [CrossRef]
8. Huang, T.H.; Hsieh, C.C.; Kuo, L.M.; Chang, C.C.; Chen, C.H.; Chi, C.C.; Liu, C.H. Malnutrition associated with an increased risk of postoperative complications following hepatectomy in patients with hepatocellular carcinoma. *HPB* **2019**, *21*, 1150–1155. [CrossRef]
9. Garth, A.K.; Newsome, C.M.; Simmance, N.; Crowe, T.C. Nutritional status, nutrition practices and post-operative complications in patients with gastrointestinal cancer. *J. Hum. Nutr. Diet.* **2010**, *23*, 393–401. [CrossRef]
10. Vashi, P.G.; Gorsuch, K.; Wan, L.; Hill, D.; Block, C.; Gupta, D. Sarcopenia supersedes subjective global assessment as a predictor of survival in colorectal cancer. *PLoS ONE* **2019**, *14*, 1–14. [CrossRef]
11. Martin, L.; Gioulbasanis, I.; Senesse, P.; Baracos, V.E. Cancer-Associated Malnutrition and CT-Defined Sarcopenia and Myosteatosis Are Endemic in Overweight and Obese Patients. *J. Parenter. Enter. Nutr.* **2020**, *44*, 227–238. [CrossRef] [PubMed]
12. Almasaudi, A.S.; McSorley, S.T.; Dolan, R.D.; Edwards, C.A.; McMillan, D.C. The relation between Malnutrition Universal Screening Tool (MUST), computed tomography–derived body composition, systemic inflammation, and clinical outcomes in patients undergoing surgery for colorectal cancer. *Am. J. Clin. Nutr.* **2019**, 1–8. [CrossRef] [PubMed]
13. Gabrielson, D.K.; Scaffidi, D.; Leung, E.; Stoyanoff, L.; Robinson, J.; Nisenbaum, R.; Brezden-Masley, C.; Darling, P.B. Use of an abridged scored patient-generated subjective global assessment (abPG-SGA) as a nutritional screening tool for cancer patients in an outpatient setting. *Nutr. Cancer* **2013**, *65*, 234–239. [CrossRef] [PubMed]
14. Abbott, J.; Teleni, L.; McKavanagh, D.; Watson, J.; McCarthy, A.L.; Isenring, E. Patient-Generated Subjective Global Assessment Short Form (PG-SGA SF) is a valid screening tool in chemotherapy outpatients. *Support. Care Cancer* **2016**. [CrossRef] [PubMed]
15. Nakanishi, R.; Oki, E.; Sasaki, S.; Hirose, K.; Jogo, T.; Edahiro, K.; Korehisa, S.; Taniguchi, D.; Kudo, K.; Kurashige, J.; et al. Sarcopenia is an independent predictor of complications after colorectal cancer surgery. *Surg. Today* **2018**, *48*, 151–157. [CrossRef] [PubMed]
16. Aro, R.; Mäkäräinen-Uhlbäck, E.; Ämmälä, N.; Rautio, T.; Ohtonen, P.; Saarnio, J.; Meriläinen, S. The impact of sarcopenia and myosteatosis on postoperative outcomes and 5-year survival in curatively operated colorectal cancer patients–A retrospective register study. *Eur. J. Surg. Oncol.* **2020**. [CrossRef]

17. Martin, L.; Hopkins, J.; Malietzis, G.; Jenkins, J.T.; Sawyer, M.B.; Brisebois, R.; MacLean, A.; Nelson, G.; Gramlich, L.; Baracos, V.E. Assessment of Computed Tomography (CT)-Defined Muscle and Adipose Tissue Features in Relation to Short-Term Outcomes After Elective Surgery for Colorectal Cancer: A Multicenter Approach. *Ann. Surg. Oncol.* **2018**, *25*, 2669–2680. [CrossRef]
18. Cespedes Feliciano, E.M.; Kroenke, C.H.; Meyerhardt, J.A.; Prado, C.M.; Bradshaw, P.T.; Kwan, M.L.; Xiao, J.; Alexeeff, S.; Corley, D.; Weltzien, E.; et al. Association of Systemic Inflammation and Sarcopenia with Survival in Nonmetastatic Colorectal Cancer, Results from the C SCANS Study. *JAMA Oncol.* **2017**, *3*, e172319. [CrossRef]
19. Hopkins, J.J.; Reif, R.L.; Bigam, D.L.; Baracos, V.E.; Eurich, D.T.; Sawyer, M.B. The impact of muscle and adipose tissue on long-term survival in patients with stage I to III colorectal cancer. *Dis. Colon Rectum* **2019**, *62*, 549–560. [CrossRef]
20. Martin, L.; Birdsell, L.; MacDonald, N.; Reiman, T.; Clandinin, M.T.; McCargar, L.J.; Murphy, R.; Ghosh, S.; Sawyer, M.B.; Baracos, V.E. Cancer cachexia in the age of obesity: Skeletal muscle depletion is a powerful prognostic factor, independent of body mass index. *J. Clin. Oncol.* **2013**, *31*, 1539–1547. [CrossRef]
21. Prado, C.M.; Lieffers, J.R.; McCargar, L.J.; Reiman, T.; Sawyer, M.B.; Martin, L.; Baracos, V.E. Prevalence and clinical implications of sarcopenic obesity in patients with solid tumours of the respiratory and gastrointestinal tracts: A population-based study. *Lancet Oncol.* **2008**, *9*, 629–635. [CrossRef]
22. Kazemi-Bajestani, S.M.R.; Mazurak, V.C.; Baracos, V. Computed tomography-defined muscle and fat wasting are associated with cancer clinical outcomes. *Semin. Cell Dev. Biol.* **2016**, *54*, 2–10. [CrossRef] [PubMed]
23. Mourtzakis, M.; Prado, C.M.M.; Lieffers, J.R.; Reiman, T.; McCargar, L.J.; Baracos, V.E. A practical and precise approach to quantification of body composition in cancer patients using computed tomography images acquired during routine care. *Appl. Physiol. Nutr. Metab.* **2008**, *33*, 997–1006. [CrossRef] [PubMed]
24. Fischer, M.; Jevenn, A.; Hipskind, P. Evaluation of muscle and fat loss as diagnostic criteria for malnutrition. *Nutr. Clin. Pract.* **2015**, *30*, 239–248. [CrossRef] [PubMed]
25. Thoresen, L.; Frykholm, G.; Lydersen, S.; Ulveland, H.; Baracos, V.; Prado, C.M.M.; Birdsell, L.; Falkmer, U. Nutritional status, cachexia and survival in patients with advanced colorectal carcinoma. Different assessment criteria for nutritional status provide unequal results. *Clin. Nutr.* **2013**. [CrossRef] [PubMed]
26. Moctezuma-Velazquez, C.; Ebadi, M.; Bhanji, R.A.; Stirnimann, G.; Tandon, P.; Montano-Loza, A.J. Limited performance of subjective global assessment compared to computed tomography-determined sarcopenia in predicting adverse clinical outcomes in patients with cirrhosis. *Clin. Nutr.* **2019**. [CrossRef] [PubMed]
27. Malmstrom, T.K.; Miller, D.K.; Simonsick, E.M.; Ferrucci, L.; Morley, J.E. SARC-F: A symptom score to predict persons with sarcopenia at risk for poor functional outcomes. *J. Cachexia Sarcopenia Muscle* **2016**, *7*, 28–36. [CrossRef]
28. Ní Bhuachalla, É.B.; Daly, L.E.; Power, D.G.; Cushen, S.J.; MacEneaney, P.; Ryan, A.M. Computed tomography diagnosed cachexia and sarcopenia in 725 oncology patients: Is nutritional screening capturing hidden malnutrition? *J. Cachexia Sarcopenia Muscle* **2018**, *9*, 295–305. [CrossRef]
29. Shen, W.; Punyanitya, M.; Wang, Z.M.; Gallagher, D.; St.-Onge, M.P.; Albu, J.; Heymsfield, S.B.; Heshka, S. Total body skeletal muscle and adipose tissue volumes: Estimation from a single abdominal cross-sectional image. *J. Appl. Physiol.* **2004**, *97*, 2333–2338. [CrossRef]
30. Gillis, C.; Nguyen, T.H.; Liberman, A.S.; Carli, F. Nutrition adequacy in enhanced recovery after surgery: A single academic center experience. *Nutr. Clin. Pract.* **2015**, *30*, 414–419. [CrossRef]
31. Gillis, C.; Carli, F. Promoting perioperative metabolic and nutritional care. *Anesthesiology* **2015**, *123*, 1455–1472. [CrossRef] [PubMed]
32. Weimann, A.; Braga, M.; Carli, F.; Higashiguchi, T.; Hübner, M.; Klek, S.; Laviano, A.; Ljungqvist, O.; Lobo, D.N.; Martindale, R.; et al. ESPEN guideline: Clinical nutrition in surgery. *Clin. Nutr.* **2017**, *36*, 623–650. [CrossRef] [PubMed]
33. Hopkins, J.J.; Reif, R.; Bigam, D.; Baracos, V.E.; Eurich, D.T.; Sawyer, M.M. Change in Skeletal Muscle Following Resection of Stage I–III Colorectal Cancer is Predictive of Poor Survival: A Cohort Study. *World J. Surg.* **2019**, *43*, 2518–2526. [CrossRef] [PubMed]
34. Xiao, J.; Caan, B.J.; Weltzien, E.; Cespedes Feliciano, E.M.; Kroenke, C.H.; Meyerhardt, J.A.; Baracos, V.E.; Kwan, M.L.; Castillo, A.L.; Prado, C.M. Associations of pre-existing co-morbidities with skeletal muscle mass and radiodensity in patients with non-metastatic colorectal cancer. *J. Cachexia Sarcopenia Muscle* **2018**, *9*, 654–663. [CrossRef]

35. Derstine, B.A.; Holcombe, S.A.; Ross, B.E.; Wang, N.C.; Su, G.L.; Wang, S.C. Skeletal muscle cutoff values for sarcopenia diagnosis using T10 to L5 measurements in a healthy US population. *Sci. Rep.* **2018**, *8*, 11369. [CrossRef]
36. Miller, J.; Wells, L.; Nwulu, U.; Currow, D.; Johnson, M.J.; Skipworth, R.J.E. Validated screening tools for the assessment of cachexia, sarcopenia, and malnutrition: A systematic review. *Am. J. Clin. Nutr.* **2018**, *108*, 1196–1208. [CrossRef]
37. Gillis, C.; Li, C.; Lee, L.; Awasthi, R.; Augustin, B.; Gamsa, A.; Liberman, A.S.; Stein, B.; Charlebois, P.; Feldman, L.S.; et al. Prehabilitation versus Rehabilitation. *Anesthesiology* **2014**, *121*, 937–947. [CrossRef]
38. Barberan-Garcia, A.; Ubré, M.; Roca, J.; Lacy, A.M.; Burgos, F.; Risco, R.; Momblán, D.; Balust, J.; Blanco, I.; Martínez-Pallí, G. Personalised Prehabilitation in High-risk Patients Undergoing Elective Major Abdominal Surgery. *Ann. Surg.* **2018**, *267*, 50–56. [CrossRef]
39. Majumdar, D.; Dale, E.; Childs, M.; Clough, S.; Loveridge, R. Initiation of physical activity prehabiliation and rehabilitation for major colorectal surgery. *Clin. Nutr. ESPEN* **2019**. [CrossRef]
40. Silver, J.K.; Baima, J. Cancer prehabilitation: An opportunity to decrease treatment-related morbidity, increase cancer treatment options, and improve physical and psychological health outcomes. *Am. J. Phys. Med. Rehabil.* **2013**, *92*, 715–727. [CrossRef]
41. Moore, J.; Merchant, Z.; Rowlinson, K.; Mcewan, K.; Evison, M.; Faulkner, G.; Sultan, J.; Mcphee, J.S.; Steele, J. European Journal of Surgical Oncology Implementing a system-wide cancer prehabilitation programme: The journey of Greater Manchester's 'Prehab4cancer'. *Eur. J. Surg. Oncol.* **2020**. [CrossRef] [PubMed]

MDPI
St. Alban-Anlage 66
4052 Basel
Switzerland
www.mdpi.com

Nutrients Editorial Office
E-mail: nutrients@mdpi.com
www.mdpi.com/journal/nutrients

www.ingramcontent.com/pod-product-compliance
Lightning Source LLC
Chambersburg PA
CBHW061604210926
51850CB00009B/2205
9783725805334